Technological Humanities

Jennifer L. Lieberman • Anto Mohsin
Editors

Technological Humanities

Essays in Honor of Ronald R. Kline

palgrave
macmillan

Editors
Jennifer L. Lieberman
Community College of
Allegheny County
Pittsburgh, PA, USA

Anto Mohsin
Northwestern University in Qatar
Doha, Qatar

ISBN 978-3-032-24740-7 ISBN 978-3-032-24741-4 (eBook)
https://doi.org/10.1007/978-3-032-24741-4

This Palgrave Macmillan imprint is published by the registered company Springer Nature Switzerland AG.
The registered company address is: Gewerbestrasse 11, 6330 Cham, Switzerland

This book is dedicated to Ronald R. Kline, scholar, co-author, teacher, and mentor.

Acknowledgements

This volume grew out of conversations (some long-running and others serendipitous) that reflect the kinds of intellectual communities Ronald R. Kline has spent a lifetime cultivating. It is therefore fitting that *Technological Humanities* takes shape not simply as a tribute to Ron's scholarship but as an expression of the relationships, mentorships, and shared commitments that his work has inspired. We dedicate this book to Ron with deep gratitude and admiration, in recognition of the profound impact he has had as a scholar, teacher, editor, and mentor. His insistence that technologies be understood through the lives, values, and agency of people, especially those often overlooked in dominant narratives, has shaped entire fields and continues to guide how many of us think, teach, and write.

We are especially thankful to Ron for his generosity in agreeing to be interviewed for this volume and for taking the time to reflect with us on his intellectual journey, his teaching, and his continuing engagement with questions about technology and society. That conversation—thoughtful, candid, and grounded in storytelling—captures the spirit of this book and models the kind of humanistic attention to technology that Ron has long championed. It was a privilege for us both, as editors and former students of his work, to learn from him once again in dialogue.

The idea for this collection began, fittingly, in conversation. On February 16, 2024, Anto spoke with Ray and Suzanne in Evanston, Illinois. Over coffee, he shared a long-held desire to edit a collection in honor of Ron, one that would not only celebrate his scholarship but also reflect the range of work his mentorship has made possible. Ray immediately suggested reaching out to Jenni Lieberman, noting that she, too, had been thinking along similar lines. Jenni readily accepted the invitation to collaborate, and from that moment the project became a shared endeavor in the fullest sense. The ease with which this collaboration formed speaks directly to the intellectual generosity and collegial culture Ron himself has modeled.

We extend our sincere thanks to the contributors to this volume who represent Ron's co-authors, advisees, and even "grand student": Clinton J. Andrews, Lina Moe, Jamey Wetmore, Christina Dunbar-Hester, Ray Fouché, Nathan Kapoor, Suzanne Moon, Ling-Fei Lin, Heather A. Love, and Honghong Tinn. Each of these scholars brings a distinctive voice, methodology, and set of concerns to the volume, yet all share a commitment to understanding technology as a human practice that is historically situated, politically contested, and socially meaningful. Their chapters engage Ron's work in diverse ways: by extending his insights into new empirical domains, by revisiting familiar themes through fresh analytical lenses, and by carrying forward his emphasis on users, teachers, and communities as central actors in technological change. We are grateful not only for the excellence of their contributions but also for their patience, responsiveness, and generosity throughout the editorial process.

We also thank the editorial team at Palgrave Macmillan. Rachel Ballard and Karthika Devi Ravikumar who guided this project with care and enthusiasm from its earliest proposal through the final stages of manuscript preparation. Their support made it possible for us to shape the book thoughtfully and deliberately, and their confidence in the project sustained it through the inevitable complexities of editing an interdisciplinary collection.

This project was nurtured within supportive institutional environments, and we gratefully acknowledge the Community College of Allegheny County and Northwestern University in Qatar for providing the conditions that allowed this work to unfold. Though different in context and mission, both institutions value teaching, critical inquiry, and engaged scholarship, and they have shaped how we think about the role of the humanities in technological societies.

Finally, we wish to acknowledge the broader networks of colleagues, students, and interlocutors whose informal and unrecorded conversations have influenced this book. Ron's legacy reminds us that scholarship is not only produced in monographs and articles but also in classrooms, advising meetings, conference hallways, and shared moments of intellectual curiosity. If this volume succeeds in foregrounding those human dimensions of technological inquiry, it is because it stands within a tradition of mentorship and community that Ron helped build and that continues to flourish through the work of those gathered here.

Contents

Notes on Contributors

Clinton J. Andrews directs the Center for Urban Policy Research at Rutgers University, where he is Distinguished Professor of Urban Planning and Policy Development. He was educated at Brown University and Massachusetts Institute of Technology in Engineering and Planning, and he worked previously in the private sector and at Princeton University. He performs research on how people use the built environment, addressing how technological and environmental changes affect urban life. His books include *Humble Analysis: The Practice of Joint Fact-Finding*, *Regulating Regional Power Systems* and *Industrial Ecology and Global Change*. He is a Fellow of AAAS stands for American Association for the Advancement of Science and IEEE is an abbreviation for Institute of Electrical and Electronics Engineers.

Christina Dunbar-Hester is the author of *Oil Beach: How Toxic Infrastructure Threatens Life in the Ports of Los Angeles and Beyond* (2023); *Hacking Diversity: The Politics of Inclusion in Open Technology Cultures* (2020); and *Low Power to the People: Pirates, Protest, and Politics in FM Radio Activism* (2014); as well as numerous popular and academic articles and book chapters. Dunbar-Hester received her doctoral degree in Science and Technology Studies (STS) from Cornell University. She is Professor of Communication in the University of Southern California's Annenberg School.

Rayvon Fouché holds a joint appointment as Professor of Communication Studies in the School of Communication and Professor of Journalism in the Medill School of Journalism, Media, and Integrative Marketing Communications at Northwestern University, where he also directs the Humanity and Technoscience (HAT) Lab. He received his Ph.D. in Science and Technology Studies (STS) from Cornell University. He is the author of *Black Inventors in the Age of Segregation* (2003), *Game Changer: The Technoscientific Revolution in Sports* (2017), and *Technoskepticism: Between Possibility and Refusal* (2025).

Nathan Kapoor is Assistant Professor of History at Illinois State University, where he teaches classes in the History of Science, Technology, and Medicine. His forthcoming book, *Powering Colonialism*, explores the history of the relationship between settler-colonialism and electrification in New Zealand.

Jennifer L. Lieberman is Associate Dean of Humanities, Social Science, and Mathematics at the Community College of Allegheny County in Pittsburgh, Pennsylvania. After completing her doctorate in English at the University of Illinois, Urbana-Champaign, she was awarded a National Science Foundation postdoctoral fellowship to study Science and Technology Studies (STS) at Cornell University. She is the author of Power Lines: Electricity in American Life and Letters, 1882–1952 (MIT Press, 2017), as well as over a dozen essays on the nexus of literary studies and STS, appearing in venues ranging from Feminist Encounters through Configurations and Transatlantic Currents: Essays in Honor of David E Nye.

Ling-Fei Lin received her doctoral degree in Science and Technology Studies (STS) from Cornell University in 2015, specializing in the history of Taiwan's PC design–manufacturing nexus and high-tech production. She conducted postdoctoral research at Nanyang Technological University (2016–2017) in Singapore and was a postdoc researcher and Asian Center Fellow at Harvard University (2018–2019). Lin's research interests have grown to encompass digital Egyptology since 2019. After studying ancient Egyptian languages and cultures, she is currently pursuing a dual Ph.D. in Informatics and Egyptology at Indiana University Bloomington, where she also teaches as a Ph.D. instructor or associate instructor.

Heather A. Love is Associate Professor of English at the University of Waterloo (Ontario, Canada), where she conducts interdisciplinary research and teaches courses on topics related to literature, culture, technology, health, and engineering education. She is the author of *Cybernetic Aesthetics: Modernist Networks of Information and Data* (Cambridge University Press, 2023), which was awarded the Gertrude J. Robinson Book Prize from the Canadian Communication Association and was a finalist for the British Society for Literature and Science Book Prize. Her work has appeared in journals such as *Modernism/modernity*, *New Literary History*, *Health Humanities*, and *IEEE Technology and Society Magazine*.

Lina Moe is a Ph.D. candidate in Planning and Public Policy at Rutgers University. Her research examines how institutions govern emerging technologies, from social robots to large-scale solar infrastructure. She is also a Research Fellow at the Federal Reserve Bank of St. Louis.

Anto Mohsin is Associate Professor in Residence in the Liberal Arts Program at Northwestern University in Qatar. He is an STS scholar specializing in infrastructure, energy, disaster, and environmental studies, with a focus on Indonesia in the twentieth century and a broader interest in Southeast Asia. He authored *Electrifying Indonesia: Technology and Social Justice in National Development* (University of Wisconsin Press, 2023). He earned his doctorate in STS from Cornell University and was trained in the university's Southeast Asia Program.

Suzanne Moon is Luke N. Walker Presidential Professor in the Department of History of Science, Technology, and Medicine at the University of Oklahoma. She specializes in the history of technology in Southeast Asia, especially Indonesia. Her current work is focused on agriculture, climate change, and sociotechnical imaginaries that inform adaptation. She was awarded the 2021 Da Vinci prize from the Society for the History of Technology (SHOT).

Honghong Tinn is an associate professor of the Program in the History of Science, Technology, and Medicine and the Department of Electrical and Computer Engineering at the University of Minnesota Twin Cities. She received her Ph.D. in STS from Cornell University. Her publications have appeared in *Technology and Culture*, *Osiris*, *IEEE Annals of the History of Computing*, and *East Asian Science, Technology and Society*. She is the author of *Island Tinkerers: Innovation and Transformation in the Making of Taiwan's Computing Industry* (MIT Press, 2025) and a co-author of *Computer: A History of the Information Machine*, 4th ed. (Routledge, 2023).

Jameson M. Wetmore is an associate professor and Director of the Center for Engagement and Training in Science & Society at Arizona State University. He completed his dissertation on the socio-political history of automobile safety in the US under the direction of Ron Kline in 2003 at Cornell University. He continues to explore the ways in which society shapes technology and vice versa and has run 17 versions of Science Outside the Lab since 2007. He is an Associate Director for Societal and Ethical Implications of the National Nanotechnology Coordinated Infrastructure and a co-author of Technology & Society: Building our Sociotechnical Future (MIT Press, 2009).

List of Figures

CHAPTER 1

Critical Humanism Confronts Technology

Jennifer L. Lieberman and Anto Mohsin

"Speaking to what the humanities do, I think about the different perspectives we can take on technical issues"
-- Ronald R. Kline

Introduction

As we craft this book, the terms *technology* and *humanities* conjure heated debates about how the former might impact the latter. For instance, public discourse about artificial intelligence (AI) attests that technological determinism is just as rampant a cultural myth today as it was during the founding of the History of Technology as a discipline in the 1950s. Major periodicals feature questions that illustrate this fact: *How will AI—often invoked synonymously with large language models—affect academic writing about history, philosophy, literature? Will higher education as we know it*

J. L. Lieberman
Community College of Allegheny County, Pittsburgh, PA, USA

A. Mohsin (✉)
Northwestern University in Qatar, Doha, Qatar
e-mail: anto.mohsin@northwestern.edu

J. L. Lieberman, A. Mohsin (eds.), *Technological Humanities*,
https://doi.org/10.1007/978-3-032-24741-4_1

survive the so-called AI age? Whose jobs will AI take first?[1] Although the subject (AI) seems new, the subject/verb pairing that positions technology as an autonomous actor is very old. Indeed, the idea that technology somehow shakes off its users, makers, and maintainers to become a deterministic force is actually rather dated. And that fact raises the more interesting question: why do we keep retelling this myth?

Answers to this question depend on who *we* signifies. The *we* who reinscribe determinist cultural narratives might denote technocrats: the people who profit from consumer goods and heroic narratives about them. These actors wield capital to promote specific uses, metaphors, and interpretations. Tech-billionaires, and the people who idolize them, might suggest that large language models do not make *mistakes;* they *hallucinate.* And that terminology will take off, so that even detractors will often feel inclined to invoke the same metaphors, killing them until they seem to become a matter of fact.

Yet money and power are not the only variables at play. The shared experiences of users can also perpetuate the very same myths. For instance, take a common end-user experience: predictive text. As we ping-ponged messages across the globe to one another and our contributors, the editors of this volume have been constantly compelled to accept or decline machine-generated phrases. "Dear Anto," Jenni would write, *"I hope this email finds you well,"* her email interface would blandly suggest. The words would flash on the screen and vanish as she typed right over them. Users who encounter these predictions throughout the workday may find them mildly useful or a bit of a nuisance. In either case, the residue of AI appears to manifest *on the screen* and *in the moment* rather than in the energy-consumptive hardware of a data center or server farm or in the dreadful workplace conditions of many AI workers.[2] And because of that dislocation, it can be prohibitively difficult for end-users to conceptualize the human decisions that have embedded predictive text in almost every

[1] For examples of these ubiquitous headlines, see: Kevin Roose, "Powerful A.I. Is Coming. We're Not Ready," Technology, *The New York Times*, March 14, 2025; Jack Kelly, "These Jobs Will Fall First As AI Takes Over The Workplace," Forbes, April 25, 2025; Ashanty Rosario, "I'm a High Schooler. AI Is Demolishing My Education," Technology, *The Atlantic*, September 3, 2025.

[2] See for example Kate Crawford, *Atlas of AI: Power, Politics, and the Planetary Costs of Artificial Intelligence* (Yale University Press, 2021); Callum Cant et al., *Feeding the Machine: The Hidden Human Labor Powering A.I.* (Bloomsbury Publishing, 2024); Karen Hao, *Empire of AI: Dreams and Nightmares in Sam Altman's OpenAI* (Penguin Press, 2025).

digital writing interface. Tech corporations (and policies that cater to them) have mobilized entire infrastructures to curate the impression that these tools function autonomously.

Still, unlike technocrats, users do not necessarily profit from technologically determinist narratives. Often the contrary is true. These narratives downplay the agency non-technologists have in shaping their own material reality. Technological determinism encourages users to sit back and wait for technology to *cause* progress; it obscures the fact that consumers have more agency than their purchasing power. The truth is, people drive the policy and the profits that help determine which products will endure and which will become obsolete. People tinker and modify. They craft counternarratives and raise crucial questions. When we interact with a technology we neither control nor wholly understand, determinism emerges as one way to narrate our experience. Determinism is, as Evelyn Fox Keller might say, *elicitable.*[3] It is one incomplete way of narrating the experience of large-scale technological change. But determinism is not the only—or even the most accurate or useful—story we can tell about technology. As Ronald R. Kline suggests in our epigraph, humanists can offer a variety of perspectives on technical issues. By listening to other voices, we can learn even more about how specific technologies work, how we can use them, how we can interpret their adoption in the past, and how we can make intentional and more informed choices about its applications in the future.

This collection posits critical humanism as an antidote to uncritical determinism.[4] In particular, it spotlights the power and potential in the school of thought developed by Professor Ronald R. Kline, former president of the Society for the History of Technology (SHOT) and the IEEE Society for the Social Implications of Technology, Leonardo Da Vinci Medal awardee (SHOT's highest honor), former chief editor of IEEE's *Technology & Society Magazine*, and Emeritus Professor of Science and Technology Studies (S&TS) at Cornell University. If determinists obscure stories of human agency, Kline spent his career bringing those overlooked

[3] Evelyn Fox Keller, *Refiguring Life: Metaphors of Twentieth-Century Biology* (Columbia University Press, 1996), xiii. In this passage, Keller discusses the interpretation of ova fertilization as seen through a microscope. The word *elicitable* connotes the often-elided interpretive work scientists do to create knowledge even when it seems visibly apparent; it also hints at the fact that other interpretations can be available.

[4] The chapter's title is inspired by, and pays homage to, the critical ethos of Judy Wajcman's book *Feminism Confronts Technology* (1991).

stories to light. This book, then, plays two roles: it serves as a *festschrift* that illuminates and honors the field-building work of a visionary scholar, whose words hold urgent relevance to us today. It also comprises a collection of essays that stand on their own, each modeling ways we might productively study technology and the humanities together.

Professor Ronald R. Kline—Ron, as we know him—has embodied both the role of the producer (he was an electrical engineer) and the analyst (he retired a gifted and prolific historian of technology). Perhaps because of his experience as an engineer, Ron was able to approach the history of technology as a history of *people*. Whether he was studying a luminary, like Charles P. Steinmetz, or the people who consumed and reimagined electricity in the country, he highlighted the human choices that shaped the uses and the meanings of technology. Given that this text pays homage to an important scholar while illuminating his legacy across a range of fields, we will first summarize his varied body of work before guiding readers through the intellectual arc our collection of essays follows.

Technological Humanities

Ron's influence on the ways of knowing we call the *technological humanities* cannot be boiled down to his three impressive monographs—more on this point below.[5] Still, these texts offer an elucidating place to begin. They outline the story of a teacher-scholar's intellectual life. And, as Ron says in our interview with him at the end of this volume, "I can't explain anything without a story." It is no coincidence that many historians of technology call themselves, borrowing from John M. Staudenmaier's book title, *technology's storytellers*.[6]

The first story Ron told about technology began with a field he knew quite well: electrical engineering. He was drawn to study one particular engineer, whose biography includes motifs about mathematics, corporate socialism, and the process of transforming an immigrant with physical deformities into an American icon of genius: Charles P. Steinmetz. Ron's

[5] Kline also produced dozens of articles in scholarly and popular journals, as well as an archive of teaching and mentorship that we discuss in the conclusion to this essay. See his Google Scholar list: https://scholar.google.com/citations?user=skLUa4EAAAAJ&hl=en

[6] John M. Staudenmaier, *Technology's Storytellers: Reweaving the Human Fabric* (The Society for the History of Technology and MIT Press, 1985).

book, *Steinmetz: Engineer and Socialist*,[7] recounts the life and legacy of the German-born engineer and scientist. A hybrid persona, Steinmetz wrote books and articles on both technical and social engineering. His contributions in electrical engineering helped make the discipline more scientific. As Ron says, Steinmetz's "mathematical theories of electrical machines helped create a quasi-autonomous engineering science." [8] In other words, engineers do not merely apply scientific principles; they could also develop a new scientific discipline based on their own practices.

Steinmetz chronicles not only the evolution of a brilliant electrical engineer but also the construction of a modern myth. The engineer's journey from a politically radical immigrant to a celebrated corporate scientist personifies the tensions between individual genius and institutional power, between socialism and capitalism, and between scientific rigor and public spectacle. According to Ron, Steinmetz embodied several paradoxes: he had "a big brain and little body…He was the Socialist chief engineer of one of America's largest capitalist corporations, a rugged individual employed by a huge bureaucracy, an upholder of corporate values who worked at home or at a summer camp, an inveterate smoker who spent his summers swimming and canoeing."[9] His technical contributions to alternating current, lightning arresters, and electrical theory were foundational, yet it was his eccentric persona and symbolic value that propelled him into the cultural spotlight as "the Wizard of Schenectady."[10] To reconcile the contrasts in his own life, Steinmetz proposed a theory of corporate socialism, imagining that corporations could raise the standard of living for everyone in a more organized and expert-driven way than labor unions.[11]

In a sense, *Steinmetz* allowed Ron to focus the critical methodologies he learned in graduate school on his own expertise. As a trained engineer who worked on a nuclear submarine's computer control system for General Electric (GE), he was naturally drawn to study engineers. Yet as he continued to co-construct the nascent discipline of STS at Cornell

[7] Ronald R. Kline, *Steinmetz: Engineer and Socialist* (Johns Hopkins University Press, 1992).

[8] Ronald R. Kline, "Humans and Machines," *Technology and Culture* 58, no. 3 (2017): 836. This essay is the published form of Kline's acceptance speech for the Leonardo Da Vinci medal. No Reference.

[9] Kline, *Steinmetz*, 308.

[10] Kline, "Humans and Machines," 308.

[11] Kline, "Humans and Machines," 308.

University, he realized that engineers are not always central to stories about technology. Users matter too. Drawing on his life experience living in the Kansas countryside in the United States, Ron realized that there were stories about real technical ingenuity that could fall between the desks of the historian of technology and the social historian. He grew up watching people "make do," and his new analytical perspective reframed that work as an undervalued form of innovation.

As Ron mentions in the interview, which serves as the conclusion to this volume, his turn to his second monograph, *Consumers in the Country*, raised eyebrows at the time. Although *Steinmetz* was an unconventional public figure, he was a technologist nonetheless; it made sense for a professor with a joint appointment in the S&TS and electrical engineering departments to study him. To Ron's contemporaries, *Consumers* seemed like a break in tradition and methodology. A nonconformist who was unfazed by skepticism, Ron returned to the archives. But this time, he was not looking to complicate the public narrative about a single man. Rather, he sought to complicate the public narrative about where innovation happens altogether.

Ron is particularly adept at finding artifacts other historians might gloss over and lingering with them, listening for the stories they might convey. For example, in his introduction to *Consumers in the Country*, Ron provides three images: a scissors picture "showing the technological optimism of the progressive ideology of electricity in the 1930,"[12] a graph of urban technologies (telephone, electricity, running water, radio) adoption on the American farm,[13] and a General Electric (GE) diagram showing "Agencies influencing the farm Family. The mediating network of electrical modernization from GE's point of view in 1938."[14] He uses these figures to "explor[e] the powerful ideologies that inform the scissors picture and the GE diagram and examin[e] what the drawings leave out: the attempts to urbanize the farm before electricity and the complex interactions among technology, promoters, and farmers."[15] Ron demonstrates how such pieces of ephemera can encapsulate larger arguments about bottom-up efforts to craft alternative modernities. That is, rather than accepting

[12] Ronald R. Kline, *Consumers in the Country: Technology and Social Change in Rural America* (Johns Hopkins University Press, 2000), 4.

[13] Kline, *Consumers in the Country*, 5.

[14] Kline, *Consumers in the Country*, 17.

[15] Kline, *Consumers in the Country*, 19.

wholesale urban notions of the modern, Ron discovered how rural folk constructed their own imaginaries (their own sense of what the good life should hold) and their own material realities (their own ways of using technology).

Stitching histories together with rigorous archival study, he convincingly argues that Americans who lived in the countryside did not simply conform to modernizing attempts from the top, espoused by urban dwellers. Ultimately, *Consumers in the Country* demonstrates that rural Americans did not simply *adopt* technologies introduced to them (telephone, car, and electricity); they *adapted* them. Modernization was not a uniform process but a mosaic of local practices, shaped by class, gender, race, region, and ideology. The US Rural Electrification Administration (REA) and other promoters envisioned a suburbanized countryside, but farm families pursued their own paths, selectively integrating technologies into their lives. The result was a rural modernity that was plural, negotiated, and deeply rooted in community.

Just as *Consumers* seemed like a radical break from *Steinmetz*, Ron's most recent book, *The Cybernetics Moment*, offers a fresh methodology, perspective, and archive.[16] Indeed, the same nonconformity that attracted Ron to Steinmetz and inspired him to examine rural users led him to focus more on Norbert Wiener, against the advice of his engineering colleagues to study Claude Shannon.[17] In this book, Ron traces cybernetics and information theory from their World War II roots through their heyday in the 1950s–60s, into their crisis in the 1970s. He discusses how both cybernetics and information theory took on new shapes rather than disappearing. Although cybernetics was often promoted as a "universal science," it was never unified. Multiple disciplines (including biology, social science, engineering, philosophy) appropriated, contested, and transformed its concepts. Kline argues that this disunity is not a failure but essential to its history. As such, a recurrent theme is the shifting analogy: humans conceived as machines (imagining a brain as a computer, modeling social systems as feedback-driven machines) and machines conceived as human (AI, bionics, and cyborgs). This blurring provoked both excitement and anxiety in the historical actors Ron studied. Kline situates cybernetics in Cold War politics, in military-industrial patronage, in popular culture (science fiction, public intellectuals), in counterculture, etc. The

[16] Ronald R. Kline, *The Cybernetics Moment* (Johns Hopkins University Press, 2015).

[17] Kline, "Humans and Machines," 839.

book also reflects on what the cybernetics moment leaves us today: how our current emphasis on "information," "big data," "feedback," etc., carries forward both enablers and risks. Kline suggests we need to recover "paths not taken," contestations forgotten, to better understand current discourses.

The scope of these texts is impressive and diverse, and they speak to Ron's expansive curiosity and careful scholarship. In terms of content, these monographs may appear unrelated. But each of his books and essays are united by their careful attention to marginalized voices and artifacts. Even when studying prominent figures such as Weiner or Steinmetz, Ron doesn't excise their technical writing and technological accomplishments from the larger stories of their humanity, their relationships, their failures, their dreams. He chooses to study the technical and the humanistic elements together, firm in the conviction that any separation between these two realms would impose artificial boundaries on complex lived experiences. User studies have become a prominent mode in STS since Ron published *Consumers in the Country*, but this principle that the humanities are essential for creating new knowledge about technology exceeds one methodology or one subfield.

It is for the above reasons that we use *technological humanities* to name a stance rather than a field: a way of knowing that insists technologies are made, used, maintained, modified, repaired, and taught by people. It is the habit of mind that critically questions technical issues in their broader contexts. While some of these methods fall under the aegis of STS, we emphasize here that the humanistic study of technology did not become a silo. It became an incubator.

Insights from Ron and from scions in his intellectual tradition have demonstrated the power of studying varied standpoints on technical matters. These scholars have examined how knowledge and artifacts are designed, developed, transferred, adopted, used, or rejected by knowledge workers, technically skilled people, invisible technicians, marketers, and all kinds of users. This type of scholarship has shown how the development of science and technology is historically contingent and culturally situated, as well as how they can be gendered, racialized, hierarchical, ideologically influenced, and politically motivated. For this reason, too, most of the essays in this collection can be read as technological stories about how societies absorb, resist, adapt, and reconfigure technologies, often under constraint.

Navigating the Volume

When Ron shifted from studying technologists to attending to the innovative practices of rural folk, his methodological turn was considered radical. The essays we include in this collection demonstrate how that innovative turn has bred a healthy variety of approaches across disciplines, with implications reaching beyond academe. Each of the following chapters delineates stories of creative adoption and resistance; each records and amplifies voices that challenge dominant narratives about technology. That is to say: each builds on the tradition Ron began. These essays are urgent in our current moment because they remind us of the various forms of agency non-technologists have.

The first chapter in our collection, "Generative Discursive 'Disunity': Norbert Wiener's Disciplinary Agility as Contemporary Rhetorical Toolkit" by Heather A. Love, picks up Ron's call in *The Cybernetics Moment* to "recove[r] the contestations of the past and the paths not taken."[18] Methodologically resonant with Ron's study of Steinmetz, Love takes a figure who is well-known in the history of technology—Norbert Wiener—and examines the elements of his body of work that have been marginalized within previous scholarship. Leveraging original archival research into eclectic, unpublished writings, Love examines how Wiener's use of metaphor elucidates his interdisciplinarity. This work of technological humanities demonstrates how mutually illuminating it can be to bring the methods of literary studies to bear on our understanding of technological history. By chronicling and analyzing Wiener's unruly writing across myriad disciplines, Love demonstrates how the humanities (and her study of Wiener in particular) can offer a "toolkit" for the paths we may still yet take—both as scholars who study the intersection of technology and metaphor, and as people invested in building more equitable futures.

Ron's body of work turned from the (albeit complex and surprising) study of a "great man" to attend to voices that play a less central role in the history of technology; this volume follows the same trajectory. The second chapter in our collection is Rayvon Fouché's "The Strange Story of Automotive Collabs and Oil Embargos." Fouché writes, "One of the most important things I take from Ron is to study and embrace the way folks, not seen as the technological powerbrokers, creatively integrate technologies into their lives based on their own specific needs, wants, and

[18] Kline, *Cybernetics*, 244.

desires." In this essay, Fouché follows a personal passion of his: the automobile. This piece thoughtfully explores the intricate entanglement of this technological artifact within layered economic, social, and personal contexts. It uses the example of couture cars to illuminate how design and marketing shifts also comprise innovations that warrant the attention of technology studies scholars.

Fouché describes how socioeconomic contexts, like the oil embargo, could lead to fashion changes that allow some users to reimagine cars as part of their personal identity. Chapter 3 extends the theme of automobility into more contemporary urban contexts, examining how lightweight vehicles can inflect cultural identity at the city level. Clinton J. Andrews and Lina Moe's chapter, "Accommodating Urban Micromobility Innovations" examines tensions between private companies and public municipalities in the adoption and diffusion of lightweight vehicles within specific communities. Just as Love's chapter suggests a toolkit for scholars, Andrews and Moe's chapter offers suggestions for "adaptive governance" through a concept they define as "soft infrastructure." In other words, their chapter offers a toolkit for policymakers and scholars alike.

Expanding on Kline's argument that users can drive change, Andrews and Moe consider the dynamic tensions between three "relevant social groups" (as a Social Construction of Technology or SCOT scholar might call them): city planners, users, and private companies. Driven by different motivations, these three groups do not always work in harmony: private companies can deploy transportation programs without permits. However, users and city planners still have agency in determining how or whether to respond to these incursions. Analyzing specific case studies, Andrews and Moe demonstrate that the success and failure of car-replacement technologies cannot be understood without also considering users, planners, and various vectors of power. With an applied technological humanities approach, Andrews and Moe recommend a framework for cities to adapt to change as gracefully as possible.

Continuing the urban theme, but shifting attention away from powerbrokers, Christina Dunbar-Hester composes a biography of an artifact with shifting social meanings: the tent. The fourth chapter in this volume, "Consumers at the Coastline: The Camping Tent as Deviant Technology in Twenty-First-Century Los Angeles." Using a "material semiotics" approach, Dunbar-Hester demonstrates that the producers and marketers of polyester tents imagine that this artifact has one dominant usage: recreational camping. However, by attending closely to tent usage,

Dunbar-Hester uncovers that deviant uses of the tent abound, from its use by protestors through its proliferation in homeless settlements. Dunbar-Hester *follows* in Bruno Latour and Steve Woolgar's sense of the word; where Latour and Woolgar *follow* the "laboratory life" of scientists and engineers, technological humanists follow other unexplored trails.[19] In Dunbar-Hester's chapter, that means following a seemingly mundane artifact, the camping tent, to reveal much more complex issues and stories it connects to, including the oil industry. As in Fouché's chapter, Dunbar-Hester also considers innovations that are not strictly technical, involving usage and interpretations that break from tradition.

The subsequent chapter leaves the American setting and troubles the urban-rural dichotomy that plays a minor chord in earlier essays. Chapter 5, "Developing Dispossession: Retrofitting Settler-Colonialism at the Huntly Power Station" by Nathan Kapoor and Suzanne Moon, takes up the question of infrastructure that we saw in Andrews and Moe's chapter, this time from a postcolonial lens. Focusing on the interesting case study of Huntly Power Station, Kapoor and Moon excavate the competing imaginaries of the area. They write, "The Huntly station, as promoted by public works officials, tied modern energy production to the same settler-colonial priorities that had dominated Aotearoa-New Zealand in the past." At the same time, "For Māori and their supporters, their long history of dispossession at the hands of settlers required as much attention and correction as energy security." Huntly Power Station isn't a bygone historical example: it is an ongoing site of discussion and meaning-making between stakeholders who share different imaginaries and values.

Where Kapoor and Moon explore *intra*-national power dynamics, Ling-Fei Lin's "Manufacturing Plus: Sociotechnical Imaginaries, Interpretative Flexibility, and the Ascent of Taiwan's Pure-Play Foundry Model," the sixth chapter in our collection, examines postcolonial power dynamics on an *inter*national stage. In narrating the rise of the semiconductor industry, Lin addresses a scale that has heretofore been overlooked by the frameworks of the Social Construction of Technology (SCOT) and the sociotechnical imaginary. Lin demonstrates that Taiwan's leaders intentionally promulgated a sociotechnical imaginary she calls "manufacturing-plus"—crafting a new national identity while simultaneously challenging

[19] Bruno Latour and Steve Woolgar, *Laboratory Life: The Construction of Scientific Facts* (1986; Princeton University Press, 1986); Bruno Latour, *Science in Action: How to Follow Scientists and Engineers through Society* (Harvard University Press, 1987).

the Western notion of manufacturing as "low value." In the process of securing an intentional place in the high-tech sector of the global market, Taiwan also subverted Western constructions of where "real" innovation happens.

Paralleling Ron's body of work, the first six chapters of *Technology Humanities* remind us that agency and creativity can occur in places far from research laboratories and corporate boardrooms. It is fitting, then, that our final chapters turn to sites that are central to the life of all scholars, and yet undertheorized in scholarship: learning communities. Whether inside or outside of the conventional classroom, our final two authors remind us that educational environments can have profound effects on the social meanings of technology.

Our first several chapters demonstrate insights and interventions that technological humanities offer academics and policymakers. Chapter 7, "Teaching the History of Emerging Technologies in an Information Age" by Honghong Tinn, examines why and how these interdisciplinary studies of technology matter to students—particularly in our present moment. Tinn *follows* another seemingly mundane artifact that is often overlooked by scholars: the syllabus. Breaking down the artificial boundary between knowledge production in research and knowledge production in the classroom, Tinn demonstrates how studies of history can be transformational for students who want to understand the materialities of their present moment or to envision hopeful futures. Using her own classroom as a case study, Tinn illuminates a thesis of this collection in microcosm: *Technological Humanities* provides important counternarratives to the dominant ways of understanding technology. Like Love, Tinn shares "toolkits" with her students and her readers, demonstrating the useful interventions the different methods that comprise our subfield can make.

Chapter 8 takes up a resonant theme in a nontraditional classroom: a policy and STS program for scientists and engineers. Jameson Wetmore's "Science Outside the Lab: Ethics Training Through Science Policy Immersion" brings readers inside a workshop where Wetmore helps technologists and scientists make informed decisions. Like Andrews and Moe, Wetmore draws attention to the role of policymakers, making an interesting observation: "What I have found through previous research and confirmed through this program is that good policymakers know a fair amount of STS. They may not be trained in STS, but they are in many ways STS practitioners." In a sense, this point opens up one final black box of expertise: our own. Technological humanists are attuned to counternarratives and power dynamics that can help us build a better world, but we do not promulgate an empty-vessel approach to communication or learning. Rather, as Wetmore argues, we find that humanistic methods can create transformational opportunities to re-envision the goal of studying technology. Students in Wetmore's workshop sometimes questioned career-long assumptions about science-in-itself as a public good. Those moments of existential doubt and deep reflection can create crucial opportunities for meaningful change.

Wetmore's chapter calls to mind a quotation by Martin Luther King, Jr.: "When scientific power outruns moral power, we end up with guided missiles and misguided men."[20] And this brings us to our final chapter, "Teaching and Learning Technology Through Stories: A Conversation with Ronald R. Kline," wherein Ron describes how his own work with nuclear missiles inspired him to question the engineer's life without humanistic inquiry. In this interview, Ron discusses his intellectual journey, including the readings that inspired him, the magic of classroom conversations that bring ideas to life, and what the humanities came to mean to him as he inhabited his joint appointment in electrical engineering and a once-experimental department that has since taken root: Science and Technology Studies. He discusses his current project about artificial intelligence and we reflect, in conversation, about how much we learn in community together.

Conclusion

As Ron indicates in the line from our interview that we took as an epigraph to our introduction, the humanities offer different ways of looking at technical issues. In a sense, that line is the thesis of this collection. To return briefly to the example we started with—AI—a technological humanities perspective would begin by recognizing disunity and heterogeneity as essential to the experience of this (or any) technology. Like Ron and all of our contributors, a technological humanist would seek to amplify voices of marginalized stakeholders, whether they be AI workers, underrepresented groups in the data, or users. Positioned this way, our volume offers humane, historically grounded studies of contemporary

[20] Martin Luther King, Jr., "SPEECH: The Three Evils of Society, 1967," Black Agenda Report, January 17, 2024 https://www.blackagendareport.com/speech-three-evils-society-martin-luther-king-jr-1967

technoscience that counter the dominant narratives of "the AI age" without ignoring its inventiveness and possibilities.

Yet this collection cannot be reduced to a study of one specific technology, which seems particularly timely. It offers a suite of epistemologies—toolkits, to use Love and Tinn's metaphor—for interrogating the materialities of the past and present and for recognizing the agency we all have in co-constructing the future. Technocrats may construct narratives about powerful technologies and powerless consumers, but these stakeholders are not the only "humans in the loop," as it were. Agency is diffusive, unruly, and volatile, which means technology's stories will never be singular. Recognizing this fact, technological humanists seek perspective and knowledge in nontraditional places, knowing that the narratives they uncover can be as salubrious as they are illuminating.

With that emphasis on belying tradition in mind, we were gratified to discover a throughline in this volume that we did not anticipate in our original call for papers: teaching and mentorship. The last two essays in our volume spotlight spaces of learning as important sites of knowledge production and transformational change. But mentorship inheres as a naturally emergent theme throughout—sometimes explicitly, as in Fouché's essay, sometimes implicitly, as in Dunbar-Hester's. And this happy accident warrants some reflection in and of itself.

The genre of the *festschrift* already does some work to spotlight the relationships that are typically relegated to bibliographies and acknowledgments sections. By coalescing around a scholar's legacy, books in this genre do some work to humanize the practice of knowledge production. However, even *festschrifts* tend to conform to the academic norms. They characteristically emphasize how one person's published works shape fields and subfields. But like the scholar who focuses on only the technology and not the human relationships that give it meaning, texts that exclusively emphasize conventional research only tell a part of the story.

Each scholar's influence goes beyond their published work and extends to the myriad forms of mentorship that happen off the page. Thus, while this book honors a scholar of renown whose work has left an indelible impression on the History of Technology and STS as fields, it also goes beyond the typical citation trail to acknowledge the way Ron's influence comes to life in the ephemeral, invisible archive of teaching. Ron's writing has inspired scores of citations, including those throughout our collection. But Ron's teaching, mentorship, and camaraderie have shaped our careers—our lives—more fully than any publication could capture.

When mentorship emerged as a theme throughout our essays, we reflected on how the *festschrift* as a genre was a bit like Steinmetz's biography. This genre seems to honor a single luminary, but it also recognizes that no scholar comes into greatness alone. *Festschrifts* acknowledge the intellectual inheritances which the honored scholar has brought into their work, while putting on display the ongoing conversations that take the scholar's work in previously unanticipated directions. Drawing more attention to teaching and mentorship seems an apt way to extend this genre and to honor Ron, specifically.

Indeed, teaching is often a silent partner in knowledge creation.[21] In our interview with Ron, he mentions how he taught a class on Cold War science while brainstorming his *Cybernetics* book, and how the notes and conversations from that course have come back to inspire his emerging study of AI. This practice is common across academe; many teacher-scholars teach the texts they plan to write about. And lively conversations about those texts create an unrecoverable and invisible archive that inspires professors and students alike. This influence is vibrant and dynamic, calling to mind a quote from Octavia Butler's *Parable of the Sower:* "All that you touch / You Change. All that you Change / Changes you."[22] Ron describes this process as *stretching* each other: a powerful feeling of growing together in intellectual community.

In this sense, the *technological humanities* stance is also collective. It concerns how we answer our most pressing questions concerning technology in conversation with each other. Ron's pedagogy treated his students and mentees as intellectual equals, just as his research acknowledged that nontraditional innovators were the equals of engineers. One of the things we can all learn from Ron is to recognize how much we stand to gain when we listen even to quiet voices, and when we nourish community more than ego. At its core, therefore, this volume argues that the humanities offer indispensable ways of understanding technology by embracing disunity, amplifying marginalized voices, and attending to the human relationships— mentorship, collaboration, and community—that make knowledge possible. In honoring Ron, we honor these commitments. And

[21] For more on this idea of teaching as an overlooked archive, see, Rachel Sagner Buurma and Laura Heffernan, *The Teaching Archive: A New History for Literary Study* (University of Chicago Press, 2020).

[22] Olivia E. Butler, *Parable of the Sower* (Four Walls Eight Windows, 1993). This quotation doesn't appear only once; it is the opening of the novel and a central tenet of the religion/philosophy Earthseed that Butler's protagonist develops throughout the novel.

in taking up his toolkits, we remind ourselves that technological futures are not written by machines. People are constantly tinkering with technology, reimagining its social meanings and possible uses. Our choices and our stories matter.

Still, in keeping with Ron's real legacy—the tangible one that remains on the printed page and the harder-to-trace one that we feel as we bring his lessons with us into the world—we also hope this collection will draw attention beyond the impact of one (admittedly great) scholar. We invite others to make-visible the ephemeral and often intangible places where meaning-making happens. To our audience, we give the following invitation: when you read the essays that follow and tangle them in your own writing, assign them in your classes, and discuss them with your colleagues, remember that *all* of those human connections matter. In those connections we have the most agency of all. Because when we truly understand the value of multiple perspectives in shaping our stories about technology, we can better rise to the moment in times of seeming technological disruption to build with intention the future we aspire to together.

Bibliography

Butler, Olivia E. *Parable of the Sower*. Four Walls Eight Windows, 1993.

Buurma, Rachel Sagner, and Laura Heffernan. *The Teaching Archive: A New History for Literary Study*. University of Chicago Press, 2020.

Cant, Callum, James Muldoon, and Mark Graham. *Feeding the Machine: The Hidden Human Labor Powering A.I.* Bloomsbury Publishing, 2024.

Crawford, Kate. *Atlas of AI: Power, Politics, and the Planetary Costs of Artificial Intelligence*. Yale University Press, 2021.

King, Martin Luther, Jr. "SPEECH: The Three Evils of Society, 1967." Black Agenda Report, January 17, 2024. https://www.blackagendareport.com/speech-three-evils-society-martin-luther-king-jr-1967

Hao, Karen. *Empire of AI: Dreams and Nightmares in Sam Altman's OpenAI*. Penguin Press, 2025.

Keller, Evelyn Fox. *Refiguring Life: Metaphors of Twentieth-Century Biology*. Columbia University Press, 1996.

Kelly, Jack. "These Jobs Will Fall First As AI Takes Over The Workplace." *Forbes*, April 25, 2025.

Kline, Ronald R. *Consumers in the Country: Technology and Social Change in Rural America*. Johns Hopkins University Press, 2000.

Kline, Ronald R. "Humans and Machines." *Technology and Culture* 58, no. 3 (2017): 835–45.

Kline, Ronald R. *Steinmetz: Engineer and Socialist.* Johns Hopkins University Press, 1992.
Kline, Ronald R. *The Cybernetics Moment: Or, Why We Call Our Age the Information Age.* Johns Hopkins University Press, 2015.
Latour, Bruno. *Science in Action: How to Follow Scientists and Engineers through Society.* Harvard University Press, 1987.
Latour, Bruno, and Steve Woolgar. *Laboratory Life: The Construction of Scientific Facts.* 1986; Princeton University Press, 1986.
Roose, Kevin. "Powerful A.I. Is Coming. We're Not Ready." *The New York Times*, March 14, 2025.
Rosario, Ashanty. "I'm a High Schooler. AI Is Demolishing My Education." *The Atlantic*, September 3, 2025.
Staudenmaier, John M. *Technology's Storytellers: Reweaving the Human Fabric.* The Society for the History of Technology and MIT Press, 1985.
Wajcman, Judy. *Feminism Confronts Technology.* Pennsylvania State University, 1991.

CHAPTER 2

Generative Discursive "Disunity": Nobert Wiener's Disciplinary Agility as Contemporary Rhetorical Toolkit

Heather A. Love

Introduction

In his 2015 monograph, *The Cybernetics Moment: Or, Why We Call Our Age The Information Age*, Ron Kline charts the paired arcs of "cybernetics" and "information" as touchpoints for technological development and innovation during the twentieth century.[1] I had the pleasure of reviewing the book in 2016, where I praised how Kline's "expansive reach ... traverses fields ranging from technical experiments to philosophical debates and from CIA advisory boards to science-fiction cyborgs"; and I described

This chapter draws on research supported by the Social Sciences and Humanities Research Council of Canada.

[1] Ronald R. Kline, *The Cybernetics Moment: Or, Why We Call Our Age the Information Age* (Johns Hopkins University Press, 2015).

H. A. Love (✉)
University of Waterloo, Waterloo, ON, Canada
e-mail: halove@uwaterloo.ca

J. L. Lieberman, A. Mohsin (eds.), *Technological Humanities*,
https://doi.org/10.1007/978-3-032-24741-4_2

how his narrative "charts the events, discoveries, and interactions that first enabled a cybernetics paradigm to emerge, then fueled its heyday, and finally led to the disappearance of the term 'cybernetics' from popular discourse as 'the information age' became the defining label of late twentieth-century society."[2] While obviously rooted in the approaches and methodologies of the historian, Kline's argument has implications that are not merely historical in scope. Indeed, he proposes that the cultural narrative of cybernetics offers important lessons to twenty-first-century readers who seek ethical ways to navigate today's complex socio-technical systems. Different people and disciplines mobilized the term "cybernetics" at different moments for very different disciplinary purposes, political ends, and cultural resonance; the collection of examples Kline gathers to illustrate these claims prompts us to be skeptical of tidy, unified historical narratives of technological progress and development. And despite cybernetics' "decline" in the second half of the twentieth century (as it was "supplanted" by the idea of information as a "guiding metaphor"[3]), it also eventually experienced a "revival and reinvention" that calls into question "the inevitability of the information age" as a dominant present-day paradigm.[4] Kline's focus on the relevance of contingency—and even failure—within cybernetics' historical arc offers a broader lesson about the value of "recovering the contestations of the past and the paths not taken."[5]

As a scholar of modernist literature who has studied how experimental artistic production intersected with emerging cybernetics discourse during the years leading up to World War II,[6] I share Kline's interest in the wide-ranging interdisciplinary reach and historical lineage of the cybernetics paradigm. One of the things I most appreciate about his approach to the history of mid-twentieth-century cybernetics (and the transition that subsequently occurred within our broader cultural imaginary toward conceptions of the "information age") is that his narrative does not focus solely—or even primarily—on the technical aspects of cybernetics and information theory. Rather, he draws readers' attention to these

[2] Heather A. Love, "The Cybernetics Moment: Or Why We Call Our Age the Information Age by Ronald R. Klein (Review)," Reviews, *Configurations* (Baltimore, United States) 25, no. 4 (2017): 557–58. 10.1353/con.2017.0035.

[3] Kline, *The Cybernetics Moment: Or, Why We Call Our Age the Information Age*, 60.

[4] Kline, *The Cybernetics Moment: Or, Why We Call Our Age the Information Age*, 180, 204.

[5] Kline, *The Cybernetics Moment: Or, Why We Call Our Age the Information Age*, 244.

[6] See Heather A. Love, *Cybernetic Aesthetics: Modernist Networks of Information and Data* (Cambridge University Press, 2023). 10.1017/9781009387446.

disciplines' rhetorical and discursive dimensions. Kline invites us to consider how and why cybernetics theories circulated among varied discourse networks; and to parse the significance of the specific word choices, images, and at-times figurative language that writers use to present their ideas to both colleagues and a broader public readership. I also take inspiration from the idea of trans-historical relevance that permeates Kline's book. Present-day scholars of technology and culture have much to learn, he demonstrates, from past authors who explored the social dimensions of technological change, regardless of whether those authors' ideas, suggestions, and insights became part of our dominant frameworks.

In the spirit of Kline's invitation to explore "paths not taken" in cybernetics history, and also in light of his assessment of cybernetics as a discourse characterized by rhetorical "disunity,"[7] this essay presents a close reading of an eclectic set of writings by the "father of cybernetics," MIT mathematician Norbert Wiener.[8] The larger archive from which I draw is a set of texts I gathered at the MIT Special Collections library in February 2020, mere weeks before COVID-19 travel restrictions hit. Primarily targeting Wiener's lesser-known and unpublished writings on topics such as language, philosophy, and social values, I assembled a collection of diverse print artifacts (some in draft form, some simply hand-written, and a few published in scholarly and public venues) that offers insight into how Wiener's thinking developed as he explored topics ranging from automated factories to janitorial staff at research institutions, and from medicine's blend of art and science to the status of analogy as a tool of communication across contexts. Writing this essay has been, in many ways, a welcome opportunity to explore one of the "paths not [yet] taken" within my own research trajectory, since the exigencies of pandemic-era disruptions prevented me from immediately engaging with that archive as

[7] Scholars such as Céline Lafontaine and Andrew Pickering echo this assessment, noting that "no unified definition of cybernetics has been able to impose itself" (Céline Lafontaine, "The Cybernetic Matrix of 'French Theory'," *Theory, Culture & Society* 24, no. 5 (2007): 28. 10.1177/0263276407084637) and that "there are so many stories [of cybernetics] to be told … [that o]ne can almost say that everyone can have their own history of cybernetics" Andrew Pickering, *The Cybernetic Brain: Sketches of Another Future* (University of Chicago Press, 2011), 3, https://press.uchicago.edu/ucp/books/book/chicago/C/bo8169881.html.).

[8] Conway and Siegelman ascribe this moniker to Wiener in their biography (Flo Conway and Jim Siegelman, *Dark Hero of the Information Age* (Basic Books, 2009), https://www.hachettebookgroup.com/titles/flo-conway/dark-hero-of-the-information-age/9780786735617/?lens=basic-books).

fully as I had hoped to. In the analysis that follows, I mobilize the literary studies method of "close reading" (a term that, like cybernetics, has been defined far too many times to count)[9] to tease out some of the rhetorical and figurative strategies Wiener incorporates into his work from around 1948–1950. These are the years when the two monographs that brought him fame were published (*Cybernetics* [1948] and *The Human Use of Human Beings* [1950]);[10] During this time, he was therefore working to disseminate and promote the broader cybernetics paradigm while simultaneously struggling to grapple with and communicate what he saw as the dire implications of post-WWII/early Cold War technological innovation.

Wiener is reputed for his conviction in the social responsibility of technologists,[11] and my readings here track the contexts within which—and rhetorical strategies through which—he makes that perspective clear. His writing from the immediate post-war years is clearly occupied with the intense technological change and disruption that WWII military developments had ushered in, as well as the communicational challenges and geopolitical turmoil of the early Cold War era. A famously idiosyncratic personality, Wiener was known for taking staunch (and often controversial) ethical stances against the use of technology for violent or unjust ends. He also pursued complex projects that did not always come to frui-

[9] For recent work on the history of, and the multiple definition(s) and practice(s) that comprise, "close reading" in the field of literary studies, see John Guillory, *On Close Reading*, with Scott Newstok (University of Chicago Press, 2025), https://press.uchicago.edu/ucp/books/book/chicago/O/bo239363263.html as well as Scott Newstock's accompanying annotated bibliography (available online and introduced in brief in Scott Newstok, "A Century of 'Close Reading,'" *Public Humanities* 1 (January 2025): e90, 10.1017/pub.2025.13). One succinct definition, drawn somewhat at random from the list of "Resources" offered on Newstock's website, presents close reading as "the process by which a strong reader intellectually processes a small part of a text in order to convert that small part of the text into evidence that supports an interpretation of the larger text" (Amanda Berry, "Close Reading: What, How, and Why"). Guillory himself posits that "transmitting close reading ... is much more likely to succeed if [it] is *modeled*," so I hope that the analyses I present in this essay offer an illustrative model to readers from outside the literary studies discipline.

[10] Norbert Wiener, *Cybernetics: Or Control and Communication in the Animal and Machine* (Wiley, 1948); Norbert Wiener, *The Human Use of Human Beings: Cybernetics and Society* (Houghton Mifflin, 1950).

[11] For a well-known example demonstrating this conviction, see Wiener's 1949 letter to Walter Reuther, president of the Union of Automobile Workers, available at: https://libcom.org/history/father-cybernetics-norbert-wieners-letter-uaw-president-walter-reuther.

tion, such as the antiaircraft gun prototype he developed with Julian Bigelow during WWII and later plans for a hearing glove that received a lot of hype but never functioned accurately enough to be useful (not to mention the fact that he "fathered" a discipline that eventually, so *The Cybernetics Moment*'s argument goes, dissipated from the cultural imaginary).[12] He is therefore an ideal figure through whom to explore Kline's paired notions of "contest[ed]" technological contexts from the past and "paths not taken" in technological innovation, and to see what they have to teach us today.

The texts I analyze below illustrate Wiener's imbrication within several interrelated zones of "contestation." He is invested in exploring the status of science and scientists in society, particularly in terms of their ethical responsibility for the impacts of their work—whether those consequences are intentional/anticipated or not. And he is similarly committed to highlighting the urgency of the present moment insofar as new technological developments pose serious threats to the future of humanity. In this area, he is never afraid to identify and call out politicians and other authority figures who—to his thinking—fail to either fully grasp or sufficiently respond to those threats. I begin with an overview and affirmation of cross-disciplinary *analogy* as a central conceptual and rhetorical foundation for the cybernetics project. This principle serves as a jumping off point for expanding the scope of the cybernetic analogy to include a broader range of metaphorical, figurative, historical, and speculative tools that permeate Wiener's writing from this time. I trace the role of these tools by attending to the nuances of Wiener's linguistic choices across texts designed for diverse rhetorical situations (e.g. an oral presentation to a group of technical experts; a brief definition for inclusion in an encyclopedia-type publication; a mass media article accompanied by

[12] Greg Adamson, Ron Kline, Katina Michael, and M.G. Michael offer a brief overview of the antiaircraft project, which "did not go into operation during World War II due to complexity of its implementation" but nonetheless "led to sophisticated filtering techniques" and served as a starting point for Wiener to make connections to work related to human "nervous disorders" (see Greg Adamson et al., "Wiener's Cybernetics Legacy and the Growing Need for the Interdisciplinary Approach [Scanning Our Past]," *Proceedings of the IEEE* 103, no. 11 (2015): 2208–14, 10.1109/JPROC.2015.2483958). For a detailed account of Wiener's engagement with the hearing glove projects, see Mara Mills, "On Disability and Cybernetics: Helen Keller, Norbert Wiener, and the Hearing Glove," *Differences* 22, nos. 2–3 (2011): 74–111, 10.1215/10407391-1428852.

numerous images designed to appeal to a broad public readership; and more). I ultimately argue that the various strategies he employs are unified in their efforts to cultivate public awareness of and generate ethical responses to technology's rapidly increasing destructive potential. In doing so, I demonstrate how his shifting status as a researcher, theorist, and philosopher who collaborated and communicated widely across many academic disciplines and popular venues serves as a generative model for twenty-first-century researchers also working at a time of urgent communicational challenges and disruptive technological change.

"Exactly the Same Principle": Cybernetics' Cross-Disciplinary Analogies

The discursive strategies that Wiener employs to advance his position on these topics emerge from his richly interdisciplinary background and his embrace of analogy as a key tenet of cybernetic thinking. Before turning to Wiener's textual exploration of this idea, a brief foray into the *Oxford English Dictionary* (*OED*) offers instructive terminological grounding. According to the *OED*, an "analogy" can be either an active and instructive attempt to bring together disparate entities (i.e. "a comparison made between one thing and another for the purpose of explanation or clarification") or a sense of an intrinsically existing "correspondence between two things," which implies a more ontological "parallelism" or "equivalence."[13] These differently inflected senses of the word are both significant to Wiener's cybernetics project. In his academic publications, lectures, and media writing about cybernetics, Wiener actively makes a repeated argument about how the central concept underpinning this interdisciplinary field—i.e. communication—already exists in analogous forms across scientific disciplines such as physics, engineering, physiology, and sociology. Indeed, a descriptive biographical caption that appears below a photo of Wiener in a 1950 Life Magazine article indicates that this type of cross-disciplinary analogy is what his work was perhaps most known for advancing:

[13] "analogy (n.)," *Oxford English Dictionary*, September 2025, 10.1093/OED/1059313579.

> His prolonged studies of the striking analogies between the control systems in animal bodies and those in complex machines became the basis of his newly created cybernetics, a science of communications. Wiener has successfully applied cybernetic theory to problems of radar and naval gunnery and the "electronic brain" calculating machines, finds them helpful in understanding such unmathematical seeming subjects as law and medicine.[14]

A conviction in these analogical connections allows Wiener to propose that "exactly the same principle" can be found at play across vastly disparate fields of intellectual inquiry.[15] He thus operates on the premise of already-existing "correspondences" that link seemingly disparate disciplines (i.e. the *OED*'s ontological equivalence definition); however, because he is proposing correspondences that may seem counterintuitive to his interlocutors, he must take an active role in articulating those connections (i.e. the *OED*'s definition that emphasizes *making* comparisons). By setting the various disciplinary frameworks side-by-side and establishing their parallels, a type of "argument by proximity" emerges as a key strategy for convincing his audiences that analogical moves hold both validity and significance.[16]

To stick with definitions for a little longer, the standard cross-disciplinary analogies mentioned in the *Life Magazine* caption above, which form the foundation for the cybernetics paradigm, appear in slightly modulated ways across Wiener's *own* definitional writing. A lecture titled "Cybernetics"

[14] "How U.S. Cities Can Prepare for Atomic War (Cities that Survive the Bomb)," by K. Deutch, G. de Santillana, and Wiener, 1950, cuid16527, Box: 29A, Folder: 638. Norbert Wiener papers, MC-0022. Massachusetts Institute of Technology. Libraries. Department of Distinctive Collections, 82 (hereafter cited as Deutch, de Santillana, and Wiener, "How U.S. Cities Can Prepare," Norbert Wiener papers).

[15] Cybernetics, lecture, 1948, cuid16486, Box: 28A, Folder: 576. Norbert Wiener papers, MC-0022. Massachusetts Institute of Technology. Libraries. Department of Distinctive Collections, 8 (hereafter cited as Cybernetics lecture, 1948, Norbert Wiener papers). As something of a side note, it's interesting to see that even within the *OED*'s definitions, the interdisciplinary relevance and resonance of the term is clear: its "Technical Uses" encompass specific denotations within linguistics, philosophy, logic, zoology, and botany.

[16] In a Sept 1950 manuscript titled "The Nature of Analogy," Wiener acknowledges that analogy, as a strategy for argumentation, frequently faces critique: "almost always there arises some critic who says, 'But this is only analogy and represents a vicious sort of reasoning'" ("The Nature of Analogy," 1950, cuid16532, Box: 29B, Folder: 655. Norbert Wiener papers, MC-0022. Massachusetts Institute of Technology. Libraries. Department of Distinctive Collections, 1). It therefore becomes important to persuasively establish the *validity* of an analogical comparison before its *significance* can be recognized.

that he delivered in March 1948 to the Institute of Radio Engineers in New York and a brief 1950 entry on "Entropy and Information" provide illustrative examples.[17] Early on, the lecture proposes connections between communication engineering and "the human nervous system," pointing to their shared interest in "transmit[ting] messages."[18] Both "the modern computing machine" and "the human nervous system," he elaborates later in the talk, are "a series of switches which have two states, on and off," and they both also incorporate feedback loops that allow "the output of one stage" to become "part of the input of the next stage" in a chain of reactions.[19] The analogies between these diverse fields of inquiry are articulated as essential to cybernetics, which he describes as "a subject … that is partly physiology and partly engineering" since it involves the "combined forces" of both engineers and biologists.[20] Cybernetics researchers working across these fields, Wiener claims (making reference to the Macy Conferences),[21] have developed a notion of their disciplines' "essential unity" of purpose/process.[22] In "Entropy and Information," Wiener replicates this cross-disciplinary analogical framework to again suture disparate systems together, though this short text illustrates how those connections exist within a different set of content and contexts. Here, Wiener focuses on the concept of "the Maxwell demon," noting that it "gives at least one way for comparing entropy of light with mechanical entropy."[23] His writing thus invites us to consider analogical relationships between communication theory terminology (information, entropy, transmission) and a range of diverse biological systems such as "photosynthesis," "enzyme action[,] and similar biological catalytic phenomena."[24]

[17] "Entropy and Information," page proof, 1950, cuid16526, Box: 29A, Folder: 637. Norbert Wiener papers, MC-0022. Massachusetts Institute of Technology. Libraries. Department of Distinctive Collections, 2 (hereafter cited as "Entropy and Information," page proof, Norbert Wiener Papers).

[18] Cybernetics lecture, 1948, Norbert Wiener papers, 2.

[19] Cybernetics lecture, 1948, Norbert Wiener papers, 7.

[20] Cybernetics lecture, 1948, Norbert Wiener papers, 8.

[21] Kline's *The Cybernetics Moment* opens with a helpful account of the Macy Conferences. For the complete transactions, see Claus Pias, ed., *Cybernetics: The Macy Conferences 1946–1953. The Complete Transactions* (DIAPHANES, 2016), https://press.uchicago.edu/ucp/books/book/distributed/C/bo23348570.html.

[22] Cybernetics lecture, 1948, Norbert Wiener papers, 9.

[23] "Entropy and Information," page proof, Norbert Wiener Papers.

[24] "Entropy and Information," page proof, Norbert Wiener Papers.

When Wiener moves from discussions rooted in the hard sciences to more sociological contexts, his cross-disciplinary analogies continue, though their purpose often suggests a greater sense of *urgency*. In the final passage of the "Cybernetics" lecture, for instance, the dire implications of mechanization—which he sees as an analogical extension of the structures that exist in the human nervous system and engineering communication systems—for human labor come into focus. Explicitly labeling automation as "catastrophic," he implores his listeners to recognize that "we are facing problems from this new industrial revolution of the robot quite as serious as those we are facing from the atom bomb."[25] This rhetorical appeal to urgency likewise characterizes his more public-facing collaborative 1950 *Life Magazine* article on "How U.S. Cities Can Prepare for Atomic War," which expands the cybernetic analogies into the realm of urban infrastructure by positing that "a city is primarily a communications center, serving the same purpose as a nerve center in the body."[26] Throughout these arguments, we see Wiener drawing together analogy's dual registers as he, on the one hand, actively makes comparisons designed for rhetorical appeal, and on the other, pitches his statements as affirmations of already-existing equivalences. Furthermore, the conjunctions we see in these last two examples—automated factories are *quite AS serious AS* atomic bombs; cities serve *the same purpose AS* physiological nerve centers—signal an important feature of his linguistic style, namely, his belief in the rhetorical power of simile and metaphor.

[25] Cybernetics lecture, 1948, Norbert Wiener papers, 10.

[26] Deutch, de Santillana, and Wiener, "How U.S. Cities Can Prepare," Norbert Wiener papers, 85. I include references to this much more widely accessible *Life Magazine* piece alongside the more obscure archival examples to illustrate how Wiener translated and mobilized a fairly consistent set of ideas and rhetorical strategies for diverse settings and audiences. Interestingly, the collaborative genesis of the ideas presented in this *Life Magazine* piece affirms the cybernetic commitment to interdisciplinary discourse: the magazine frames the article's arguments as authoritative thanks to the fact that they have been produced by a "brilliant mathematician" (Wiener) working with "the best thinking of members of M.I.T.'s history and city-planning departments" (77).

From Analogy to Metaphor: Wiener as Creative/Critical Humanist

By turning to style, simile, and metaphor, I move, in the upcoming sections, to explore the rich variety of ways Wiener deploys figurative language as a means of communicating the comparisons, parallelisms, and equivalencies that comprise the cybernetics analogy. This diversity, I posit, constitutes one instance of the "disunity" Kline encourages us to probe as we explore textual narratives from the past; in Wiener's case, it highlights how cross-disciplinary agility is essential to advancing the cybernetics project beyond scientific and technical audiences. As we have seen across a range of both explicit and more subtle examples, Wiener embraces analogy as a generative way to frame cybernetics' expansive applicability across domains such as engineering, physiology, and sociology. However, if we expand our scope of analysis, we can also locate this type of "argument-by-proximity" in places when his more humanities-leaning background comes into the fore. Evidence of this desire to incorporate humanistic knowledge appears in his references to examples drawn from literary and cultural texts, as well as from concrete historical figures, events, and contexts. Although not the focus of the technical aspects of cybernetics theory, these culturally oriented elements of Wiener's thinking permeate his writing, solidifying his status as a rhetorician as much as a scientist. They often enable him to introduce complex ideas in audience-specific ways or to persuasively bolster the technical arguments he is making about the urgency of the issues he discusses and the importance of embracing the cybernetics paradigm as an ethical (and effective) response to an ever-more-complex socio-political-environmental milieu. This latter goal is a key feature of places where Wiener integrates his own playful, pointed, or at times bitter metaphors alongside vivid imagery and speculative descriptions. In these moments, we see Wiener donning the artist's hat and adopting a provocative aesthetic rooted in creative expression that traverses diverse tonal registers.

Across all of these strategies, the analogy structure dominates. Importantly, though, it also gives way to an *ethical imperative*, since it also functions as a call for scientists and other technologists to understand—and act on—their responsibility to society. This aspect of Wiener's writing is both critical and creative. On the one hand, it often becomes most legible in the bitter tone and pointed accusations that appear throughout his writing. But on the other, it also appears in the speculative moments when he attempts to linguistically construct the future he envisions and call into being a more ethical relationship between scientists/technologists/innovators and the world they inhabit. If we think of this set of texts as presenting a "toolkit" of sorts, we can see Wiener's movement across disciplinary references and styles as part of his effort to reach a broad audience and cultivate awareness of and investment in the urgency of the problems and opportunities he has identified.

A Multitude of Metaphors: Analogy, Association, Audience

Upon reviewing Wiener's work on the lookout for moments of discursive playfulness, it becomes abundantly clear that he loves nothing better than a good metaphor. Given his commitment to analogy as a foundation for cybernetic thinking, this delight in metaphorical play is perhaps unsurprising—after all, the grammatical structure of the metaphor gives us something of the analogy in miniature (according to the *OED*, a now-obsolete meaning of the term "analogy" was, in fact, "A figure of speech involving a comparison; a simile, a metaphor"[27]). In linguistics parlance, Wiener makes skilled and strategic use of the associative connections that each metaphor's "*vehicle*" cultivates for the "*primary subject* or *tenor*" it describes.[28] To start with just one illustrative example, as a means of conveying the *tangly* problem (pun intended) that U.S. transportation infrastructure poses to disaster-preparedness planning, he and his coauthors spin out an extended metaphor between the American rail system and an elaborately woven *textile*: "U.S. railroads now *weave* across the country in an elaborate *lacework* pattern of lines, almost all of which draw together at the large cities to form *knotted* intersections (*below*). A few A-bombs could *sever these knots* and collapse the entire network. The resulting *tangle* would tie up movement of food, guns, troops and medical supplies and bring about an inevitable defeat of the U.S."[29] This specific choice of metaphor complicates railroad infrastructure's typical associations with the strength

[27] *Oxf. Engl. Dict.*, "analogy (n.)."

[28] For a full discussion of the parts of a metaphor and their relationship/function, see David Hills, "Metaphor," in *The Stanford Encyclopedia of Philosophy*, Fall 2024 Edition, August 19, 2011, https://plato.stanford.edu/archives/fall2024/entries/metaphor/.

[29] Deutch, de Santillana, and Wiener, "How U.S. Cities Can Prepare," Norbert Wiener papers, 80.

and solidity of steel, the materiality and might of engines, and the physical transport of agricultural and mineral resources, not to mention people. Instead, readers are prompted to consider how the current system instead represents a national risk due to its exposed filigree detail. The metaphor is further reinforced through the imagery that accompanies the text (referenced as "*below*" in the quotation, See Fig. 2.1), which invites readers to imagine their country's dominant city layouts through the lens of knotted threads and delicate lacework. This example illustrates how attending to the connotative dimensions of the metaphors that pepper Wiener's writing offers insight into how his arguments are structured to build interest and strategically focus reader perspectives—as well as to incorporate humor, both playful and sardonic—through specific figurative details.

One theme that is often the focus of Wiener's vivid metaphors is his sense of the precarity of scientific integrity during the post-war years. Science is personified as being on *trial* (being brought to court "to prove his right to existence"[30]). Scientific development is cast as a *recklessly speeding car* (any attempt to "avoid a catastrophe by putting on the brakes,

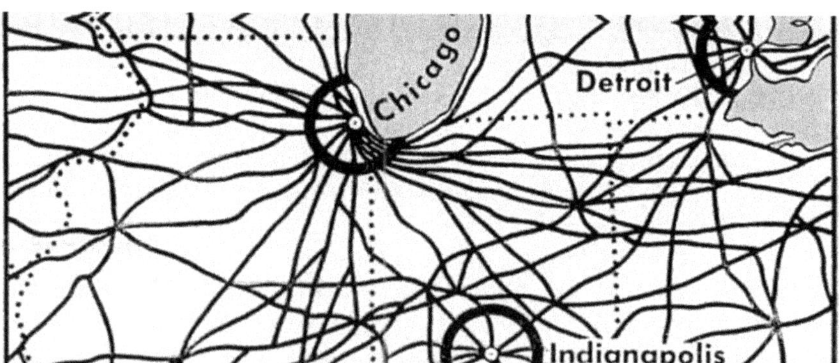

Fig. 2.1 Visual representation of the "elaborate lacework" design of current US rail infrastructure in the Midwest ("How U.S. Cities Can Prepare," 81; image credit: Anthony Sodaro and Frank Stockman, *LIFE Magazine*).

[30] "Limitations of Science," 1935, cuid16458, Box: 28A, Folder: 542. Norbert Wiener papers, MC-0022. Massachusetts Institute of Technology. Libraries. Department of Distinctive Collections, 1 (hereafter cited as "Limitations of Science," Norbert Wiener papers).

closing our eyes, and taking our hands off the wheel" is likely to result in even more chaos, given that "the road that we are driving on is slippery"[31]). Scientific accountability is a *hot potato* (which scientists "pass from hand to hand" since they fear they will "get burnt if [they] hold it too long"[32]); and without a commitment to responsibility, "we shall find that American science meets the fate of being smothered in jam."[33] This final image of "jam" calls to mind, particularly within the context of an author as interested in communication-engineering as Wiener is, the notion of a jammed communication channel. This idea provides an anchor for the figurative string of puns that appear late in the *Life Magazine* piece. Linking traffic, telephones, and human nerves, the line reads: "The traffic jams in streets and subways, the jams on the telephone circuits—jams which fray our nerves and squander our time—are evidence that our cities as they now exist are not nearly in their best possible form."[34]

In tandem with this figurative commentary on the status of *science* and the dangers it faces in the era of atomic warfare and political interference, *scientists* themselves also become richly characterized through Wiener's metaphor-riddled descriptions. He portrays the global network of scientific researchers as a tight, local community whose "connections [are] ... closer than [those] most of us have *with our neighbors*"; this community, he proposes—again with the help of a pointed metaphor—is under threat by governmental directives and economic incentives that perversely entice the scientist to become "*a kept man* of interests totally foreign to science, and totally foreign to civilization."[35]

Frequent vehicles for metaphorical resonance across Wiener's writing are medicine and health, which he deploys to illustrate his ideas about scientific expertise as well as to critique what he sees as insufficient political

[31] "Limitations of Science," Norbert Wiener papers, 2.

[32] "Too Damn Close," 1950, cuid16542, Box: 29B, Folder: 664. Norbert Wiener papers, MC-0022. Massachusetts Institute of Technology. Libraries. Department of Distinctive Collections, 13 (hereafter cited as "Too Damn Close," Norbert Wiener papers).

[33] "A Rebellious Scientist after Two Years," 1948, cuid16496, Box: 28C, Folder: 601. Norbert Wiener papers, MC-0022. Massachusetts Institute of Technology. Libraries. Department of Distinctive Collections, 339 (hereafter cited as "A Rebellious Scientist." Norbert Wiener papers).

[34] Deutch, de Santillana, and Wiener, "How U.S. Cities Can Prepare," Norbert Wiener papers, 85.

[35] "Internationalization of Science," 1948, cuid16494, Box: 28C, Folder: 599. Norbert Wiener papers, MC-0022. Massachusetts Institute of Technology. Libraries. Department of Distinctive Collections, 1, emphasis added (hereafter cited as "Internationalization of Science," Norbert Wiener papers).

action. As an example of the former, in “The Limitations of Science,” he draws on a familiar medical scenario as a simile to drive home his argument about how researchers with experience in the hard sciences can’t be expected to succeed if they shift suddenly to social scientific methods: “It is as foolish to expect that five years’ work of scientists taken out of their proper fields will make any progress worthy of the name as it would for the family of a patient dying of cancer to go into cancer research with the expectation of being of any real service to that particular patient.”[36] And in a derisive sequence of metaphorical moves condemning public commentators who downplay the threat of atomic warfare, he proclaims that their dismissals amount to an “alcoholic['s] assertion” that “Another little drink won’t hurt us,” and that their attempts to overcome risk through diplomacy are akin to “trying to treat a galloping consumption with soda-mint [sic.] tablets.”[37]

I offer these examples to provide something of the “flavor” of Wiener’s writing. As he traverses a wide range of topics and themes, he makes liberal use of metaphor, simile, and figurative language not simply to add interest and moments of illumination to his writing, but also to confront a range of different audiences with more jarring equivalencies designed to disrupt their thinking and reorient them to the urgency of the present moment. A helpful framework for approaching Wiener’s frequent integration of these figurative devices is George Lakoff and Mark Johnson’s foundational 1980 study, which explains that “*the essence of metaphor is understanding and experiencing one kind of thing in terms of another.*”[38] They emphasize that this linguistic and conceptual phenomenon both clarifies and obscures: “In allowing us to focus on one aspect of a concept ..., a metaphorical concept can keep us from focusing on other aspects of the concept that are inconsistent with that metaphor.”[39] Through this filtering effect, their examples illustrate, metaphors serve to “structure how we perceive, how we think, and what we do” as they reflect “the most fundamental values in a culture.”[40] The bulk of their theory is explicitly *not* concerned with notion of metaphor as “a device of the poetic imagination and the

[36] “Limitations of Science,” Norbert Wiener papers, 3.

[37] “Too Damn Close,” Norbert Wiener papers, 1/10.

[38] George Lackoff and Mark Johnson, *Metaphors We Live By* (The University of Chicago Press, 1980), 5.

[39] Lackoff and Johnson, *Metaphors We Live By*, 10.

[40] Lackoff and Johnson, *Metaphors We Live By*, 4/22.

rhetorical flourish"[41]—i.e. with the type of deliberately playful and provocative metaphors that populate Wiener's writing. Lakoff and Johnson instead seek to explicate how "metaphor is pervasive in everyday life, not just in language but in thought and action,"[42] and they methodically parse how the "metaphors we live by"—that is, the "systematic metaphorical concepts that structure our actions and thoughts"—are "fixed within the lexicon of English" so firmly that we don't even recognize them as metaphors.[43] In this sense, their theory aligns well with the *OED* definition of analogy that foregrounds intrinsically existing, widely recognized, and already accepted correspondences across contexts.

However, when the authors turn, in a later chapter of the book, to a brief discussion of "metaphors that are outside our conventional conceptual system," their writing offers illuminating insight into the more *active* rhetorical impulse underlying much of Wiener's work with analogy and metaphor.[44] Here, they emphasize the generative potential of "imaginative and creative" formulations: "such metaphors are capable of giving us a *new understanding* of our experience. Thus, they can give *new meaning* to our pasts, to our daily activity, and *to what we know and believe*."[45] Drawing on phrasing that evokes cybernetics discourse, they note that encountering an unconventional metaphor "can have *a feedback effect*, guiding our future actions in accordance with [how] the metaphor" resonates with our cultural frameworks and "past experiences."[46] Precisely due to their unexpected nature, imaginative metaphors can jolt us out of our quotidian habits, and if they manage to "ente[r] the conceptual system that we base our actions on," they will in fact "create a [new] reality."[47] Indeed, as this chapter powerfully puts it, "much of cultural change arises from the introduction of new metaphorical concepts and the loss of old ones."[48] These statements crystallize the goals that drive not only Wiener's explicit use of metaphor but also several more specific forms of comparative, creative reasoning that permeate and enrich his work.

[41] Lackoff and Johnson, *Metaphors We Live By*, 55.
[42] Lackoff and Johnson, *Metaphors We Live By*, 3.
[43] Lackoff and Johnson, *Metaphors We Live By*, 55.
[44] Lackoff and Johnson, *Metaphors We Live By*, 139.
[45] Lackoff and Johnson, *Metaphors We Live By*, 139, emphasis added.
[46] Lackoff and Johnson, *Metaphors We Live By*, 142, emphasis added.
[47] Lackoff and Johnson, *Metaphors We Live By*, 145.
[48] Lackoff and Johnson, *Metaphors We Live By*, 145.

Literary References: Sources of Explanation, Tools of Critique

Wiener draws on wide-ranging literary and cultural references throughout his writing, mobilizing these referents analogically and metaphorically as he strives to "create a new reality" through both explanatory techniques and pointed social critique. As an example of the former, a 1950 essay titled "Newtonian and Bergsonian Time" makes playful reference to *Alice in Wonderland* author Lewis Carroll, casting his writing as "an almost universal source of parallels and metaphors for logic in science."[49] In this particular essay, Wiener presents the reader with one of Carroll's "imaginary and humorous" poetic inventions: an animal called a "boojum," whose salient characteristic is that "he will softly and silently vanish away/ and never be heard of again."[50] The notion of a creature who immediately disappears upon contact with an observer becomes an accessible jumping off point for Wiener to explain how experimental observation operates in more technical contexts such as when studying certain types of weather ("the Wilson Cloud tract experiments"[51]) and when tracking subatomic particle collisions that can't be observed directly. These scenarios, he explains with repeated references to the boojum, require scientists to adopt a perspective from which "the history of particles is studied by the catastrophes which they produce."[52] Woven into the discussion are additional Carroll-ian shout-outs to (presumably white? perpetually late?) rabbits that add whimsy to the analogical framework as they provide a basis for Wiener's creative explanation of how Newtonian and Bergsonian theories of temporality exist in tangled tension with one another.

Opposed to this type of playfully instructive analogy, Wiener's writing takes a much darker turn when he mobilizes more ominous literary references to frame his critiques of Cold War military threats and "the position of the scientist in the modern world."[53] The atomic bomb (and, by exten-

[49] "Newtonian and Bergsonian Time," circa 1950, cuid16533, Box: 29B, Folder: 656. Norbert Wiener papers, MC-0022. Massachusetts Institute of Technology. Libraries. Department of Distinctive Collections, 4 (hereafter cited as "Newtonian and Bergsonian Time," Norbert Wiener papers).

[50] "Newtonian and Bergsonian Time," Norbert Wiener papers, 4.

[51] "Newtonian and Bergsonian Time," Norbert Wiener papers, 5.

[52] "Newtonian and Bergsonian Time," Norbert Wiener papers, 9.

[53] "Responsible Man in the Machine Age," circa 1950, cuid16535, Box: 29B, Folder: 658. Norbert Wiener papers, MC-0022. Massachusetts Institute of Technology. Libraries.

sion, its inventors) becomes lexically connected to Mary Shelley's Victor Frankenstein as a source of "monstrous progeny."[54] The mutually assured destruction that geopolitical atomic warfare makes possible—i.e. the notion that "if you use your bomb to blow us up, you will poison your own atmosphere so that the end of the United States will be the end of Russia"—is cast in terms of violently Shakespearian tragedy with Wiener's vision of a post-nuclear landscape where "the whole world will look like the stage when the curtain goes down on the last act of Hamlet.'"[55] And the complicity of twentieth-century scientists in widespread "massacre" is presented in an even more damning extended metaphor: "we stand," Wiener writes (including himself among the guilty):

> … where Macbeth stood after he had slain Duncan and before he had become aware of the sequence of tyrannies and murders into which his search for security in the face of his past conduct had led him. / We, too, with a conscience as uneasy as that of Macbeth are looking for security; not the security of the contrite heart, but the security of external protection against the consequences of our own sin. / … The times are most evil, and the hour is very late.[56]

The word games here rely, to a certain extent, on the reader's familiarity with Shakespearian plotlines for their full rhetorical appeal. Nonetheless, their basis in two of the Bard's most well-known tragic (and violent) figures ensures widespread recognition of the gravity Wiener wants to convey. Subsequent widely known Biblical references further reinforce his argument about both atomic warfare's severe consequences ("The fire of the atomic bomb will not merely burn the Bush of Life, but will consume it"[57]) and the self-condemnation he wants scientists to feel when they confront the violent implications of their work. As Wiener ironically narrates, when "obliged to declare [themselves] before the world as [sinners] … Judas went and hanged himself; Pilate spiritually washed his hands. I am afraid that I have much more confidence in the contrition of Judas."[58]

Department of Distinctive Collections, 1 (hereafter cited as "Responsible Man," Norbert Wiener papers).

[54] "Too Damn Close," Norbert Wiener papers, 2.

[55] "Too Damn Close," Norbert Wiener papers, 6.

[56] "Responsible Man," Norbert Wiener papers, 1–2.

[57] "Too Damn Close," Norbert Wiener papers, 9.

[58] "Responsible Man," Norbert Wiener papers, 2.

To look at one more figurative strategy, Wiener's ability to blend playful and pointed approaches to social commentary comes through on multiple occasions where he puns on familiar sayings and aphorisms to make points that nonetheless emphasize the urgency of the present situation. In a 1948 piece for the *Bulletin of the Atomic Scientists* where he reflects on how his thinking has progressed during the nearly two years since his famous "A Scientist Rebels" essay appeared in *The Atlantic*, he writes, "I realize, of course, that my control of my own work is not as complete as I would have it, and that there are situations in which it is scarcely worthwhile *closing the stable door after the horse has been classified.*"[59] The light tone of the "stable door" metaphor tempers the more sinister underpinning of the statement's concerns about the dampening effects of the military's careful controlled intelligence systems and document "classification" procedures. Similarly, a one-page manuscript from 1948, titled "World Politics in the Atomic Age," mobilizes darkly comic language by drawing on familiar aphorisms to issue a pointed warning and critique; amidst grim reflections on atomic weapons' capacity to cause "destruction" ranging from the "extermination of a city" through to a complete "death blow to the human race," Wiener sardonically inserts the following statement: "for the first time in history *all our eggs are in one basket* and our military men and diplomatic mentors are showing *a taste for omelettes.*"[60]

Historical Context: Exemplar, Lineage, Counterpoint

As the previous sections have demonstrated, both the overall cybernetics paradigm (with its analogy-based, cross-disciplinary frameworks) and Wiener's individual proclivity for provocative metaphors and literary/cultural referents operate according to a shared logic of argumentation that sets seemingly disparate phenomena side-by-side to illustrate their shared features. The purpose and payoff of these comparative rhetorical moves are diverse, ranging from illuminating small-scale connections and inflecting serious arguments with a humorous tone to claiming far-reaching and

[59] "A Rebellious Scientist," Norbert Wiener papers, 339, emphasis added.

[60] "World Politics in the Atomic Age," circa 1948, cuid16498, Box: 28C, Folder: 603. Norbert Wiener papers, MC-0022. Massachusetts Institute of Technology. Libraries. Department of Distinctive Collections, emphasis added (hereafter cited as "World Politics in the Atomic Age." Norbert Wiener papers).

significant alignments across disciplinary divisions. By encouraging researchers to recognize these connections, Wiener hopes to foster the type of collaborative ethos that he considers necessary to more effectively address society's most pressing problems. Another area where Wiener's "argument by analogy" draws on his humanities background is in his integration of historical references. These elements serve not only as exemplars for the various ideas he presents but also as metaphors or even extended analogies to articulate historical lineages. We see this latter strategy at play in a slightly earlier letter to the *New York Times* (1938), where he makes the case for United States having reached a "level of intricacy of internal and external affairs" so great that the country is requires to "assume imperial responsibilities"; his rhetorical method is to equate the United States' twentieth-century socio-political and economic systems to those of famously dominant historical regimes such as "Great Britain … in the late eighteenth century," "Rome at the time of Caesar," and "Vania perhaps in the Han dynasty."[61] This type of argument also appears in the 1950 "Too Damn Close" essay, where he aligns the twentieth century's status as "a very active stage in the course of developing science" with that of "the Greek period of science" and "the last quarter of the 17th century."[62] In rather broad-strokes terms of dismissal, these "active" periods are cast in contrast to "the long if unrestful sleep of the Middle Ages," "the whole of the 18th century," and the late 19th century; during this last period, he claims rather derisively, "it was the expectation of the experimental scientist that all future time would be spent in improving his results by one or two places of decimals."[63]

These examples are designed primarily to emphasize historical *continuity*, affirming the status of the United States as an imperial power and the twentieth century as a period of intense scientific innovation through historical alignments with nations, empires, and contexts reputed for epitomizing precisely these ideals. However, in other places, Wiener instead draws on historical precedent to make the opposite argument, encouraging readers to view the present state of affairs as a marked *break* with the past. Unlike "the Generals of Cromwell," who "were deep in war for

[61] New York Times letter about professionalization of US State Department, circa 1938, cuid16464, Box: 28A, Folder: 550. Norbert Wiener papers, MC-0022. Massachusetts Institute of Technology. Libraries. Department of Distinctive Collections, 1 (hereafter cited as New York Times letter, Norbert Wiener papers).

[62] "Too Damn Close," Norbert Wiener papers, 2.

[63] "Too Damn Close," Norbert Wiener papers, 2–3.

decades," for example, he claims that twentieth-century military Generals are nothing more than "amateur fighters," since the atomic weapons with which they wage war are so new that even advice from military leaders "is not worth much more than the advice of any of the rest of us."[64] One of the most powerful through-lines that emerges upon reflection on these types of passages is the idea that, at Wiener's present historical moment, technological advances are expanding the *quantitative* dimensions of what is possible so much that there has been a *qualitative* transformation that now must be confronted and addressed. Previous approaches and assumptions cannot simply be scaled up; instead, they must be radically reconfigured to address entirely new social, cultural, economic, geopolitical, and even material realities.

This sentiment appears explicitly in "World Politics in the Atomic Age," which opens with the following statement: "It has been said that since the atom bomb is only another weapon, and since the effect o[f] weapons on world politics is well-known, the atomic bomb is no great innovation, and will have merely a quantitative and not a qualitative effect. However, great quantitative changes amount to qualitative changes."[65] The basic premise that Wiener advances here is developed in a variety of ways across other texts from around this time, frequently through reference to historical points of comparison. In "Too Damn Close," for instance, he proposes that, "there is a very good case for distinguishing the atomic bomb, at least from those weapons of war which existed previous to World War Number II," and he invites readers to understand the validity of this claim through a familiar historical referent:

> War when it has been considered as legitimate has been considered as a method of imposing one's will on the other contestant, and not as a method of extermination. The *American Revolutionary War* killed many people, but neither a vital part of the population of England, nor a vital part of the population of the Colonies. The issue was whether the colonies would continue to be dominated by the policy of England, or whether they would not; and on neither side was the extermination of the people as a whole even considered. Probably, it could not have been considered at that time, because no such method of such mass extermination existed.[66]

[64] "Too Damn Close," Norbert Wiener papers, 12.
[65] "World Politics in the Atomic Age," Norbert Wiener papers.
[66] "Too Damn Close," Norbert Wiener papers, 13–14, emphasis added.

The American Revolutionary War comparison bolsters Wiener's arguments about how post-WWII geopolitical, environmental, and economic upheaval (largely shaped by disruptive technologies such as atomic weapons and—as he explores elsewhere—widespread automation[67]) should be approached as a significant crisis rather than dismissed as a set of minor societal changes. Of course, the historical referent is not alone in making this point, which Wiener also emphasizes rhetorically through repetition of the violent term "extermination."[68] But similarly structured arguments, which rely on historical points of *contrast*, appear in his descriptions of the status of science in the present—the notion that science should be "primarily a military tool" is derided by Wiener as "historically a new and false [attitude]," and he brings in references to eighteenth-century international scientific communities and the social "decay" that preceded the "fall of the Roman Empire" to make his case.[69] We also encounter an explication of historical difference in the comparisons he makes between the closed-loop assumptions of Newtonian physics and the "radically new concepts of statistical mechanics" that have emerged in modern physics.[70] Taken together, these diverse examples showcase how, in Wiener's hands, historical comparison—as a continuation of cybernetic analogy—functions in multiple ways to cultivate rhetorical urgency. It serves as a tool for establishing how the present both exists as a *continuation* of past trends and maintains an *exceptional* status with respect to prior scenarios.

[67] For a discussion of places in Wiener's writing that focus on the implications of automation for the future of work, see Heather A. Love et al., "The Future of Work in the Age of Automation: Proceedings of a Workshop on Norbert Wiener's 21st Century Legacy," *IEEE Transactions on Technology and Society*, 2024, 1–23, 10.1109/TTS.2024.3476041.

[68] Wiener's similar claims in "Responsible Man in the Machine Age," are bolstered by similar repetitions of violent terminology, the potency of which is doubly amplified in light of associative connections to not only the dropping of the atomic bombs but also the systematic mass murder and attempted eradication of Jewish (and other marginalized) people during the Nazi Holocaust: "...the scientist in the modern world ... has partaken in an undertaking which has led to *massacre* and the potential of *massacre* on an unheard of scale" / "Part of the guilt of the devotion of this new resource of the human race to *slaughter* and destruction is on his head, and he knows it" ("Responsible Man," Norbert Wiener papers, 1, emphasis added).

[69] "Internationalization of Science," Norbert Wiener papers, 1.

[70] "The Impact of Statistical Mechanics on Modern Physics and Physiology," circa 1949, cuid16504, Box: 28C, Folder: 612. Norbert Wiener papers, MC-0022. Massachusetts Institute of Technology. Libraries. Department of Distinctive Collections, 2.

Speculative World-Building

To conclude this discussion of Wiener's creative adaptations of the cybernetic analogy and metaphor frameworks, I turn to a rhetorical mode that may not seem as easily connected to those principles: speculation. Within the archive of texts I have explored here, the speculative mode appears several times, either as part of Wiener's imaginative world-building or as the subject of his analysis. Perhaps the most obvious and extended examples appear in the *Life Magazine* article, which opens with a speculative portrait of a post-nuclear-bomb scene in the United States, emphasizing horrific destruction and the paralysis that would ensue. Following the grim statement that "U.S. cities are the most destructible A-bomb targets in the world," the article launches into a description that moves from "immediate and total devastation of large areas, with casualties running into the hundreds of thousands" into volleying speculative statements about the nuclear bomb's "indirect effects" on infrastructure and populations:

> Transportation would be paralyzed, power and water cut off, food supplies destroyed. The people would abandon the city in disorganized, panic-ridden flight. Within hours local stocks of gasoline would be exhausted and every road would be blocked solidly with stalled cars. The refugees would set off on foot, pouring through suburban communities too small to feed or shelter them. Railroads, ruptured at their city terminals, could neither evacuate the homeless and injured nor bring in desperately needed supplies.[71]

Concluding this initial vision in figurative mode, "the great city" is cast "as a great explosive, triggered by the lesser explosion of the bomb itself."[72] Perhaps most overtly, the post-apocalyptic panic-scene (familiar to twenty-first-century readers/viewers of speculative genre fiction) serves in the *Life Magazine* piece as an opening foil for Wiener's alternative vision of a "plan for preparedness" that will "reckon with what happens to the people in the days and weeks which follow the dropping of the bomb."[73] However, for

[71] Deutch, de Santillana, and Wiener, "How U.S. Cities Can Prepare," Norbert Wiener papers, 77.

[72] Deutch, de Santillana, and Wiener, "How U.S. Cities Can Prepare," Norbert Wiener papers, 77.

[73] Deutch, de Santillana, and Wiener, "How U.S. Cities Can Prepare," Norbert Wiener papers, 77.

the purposes of my own analysis, the descriptions also help us see how speculation aligns with analogy and metaphor as a comparative rhetorical mode for setting side-by-side seemingly disparate contexts to focus our attention on and cultivate recognition of shared patterns. *What we are witnessing at present*, the speculative "analogy" posits (in the case of this *Life Magazine* piece, "the present" is encapsulated by the image of urban traffic jams that "clot New York's streets in peacetime"; see Fig. 2.2), *equates to a particular imagined future* (in this first portrait, the devastating "consequences of an atomic attack").[74]

Furthermore, the speculative mode, deployed to communicate the logic of "the Wiener civil defense plan," is not merely a tool for pessimistic

Fig. 2.2 Cover image for the *Life Magazine* article, accompanied by the caption "Even in peacetime New York's Fifth Avenue is paralyzed by congestion" ("How U.S. Cities Can Prepare," 77; image credit: Andreas Feininger, *LIFE Magazine*).

[74] Deutch, de Santillana, and Wiener, "How U.S. Cities Can Prepare," Norbert Wiener papers, 77.

visions for the future. Rather, it sets out to become an anticipatory source of cybernetic feedback that will cultivate responsiveness and, ideally, change the narrative—*what we are witnessing at present can ALSO equate to a more DESIRABLE imagined future, IF we act now*. The details of Wiener's vision appear on subsequent pages of the article, which outline a proposed system of carefully designed roads ("radiating like the spokes of a wheel from the center of the metropolitan area to serve as exits for the people") ringed by an "express highway" and "railroad belt" (see Fig. 2.3); "zoning regulation" that will balance supply chains, residential needs, and safety considerations; and attentiveness to both the present desires for suburban living and improved traffic flow, and the future possibility of needing to orchestrate a rapid evacuation and then feed and house the "fleeing population."[75]

In Wiener's plans, as well as the ways they are presented (i.e. side-by-side visions of a disastrous and a prosperous future that recall, structurally,

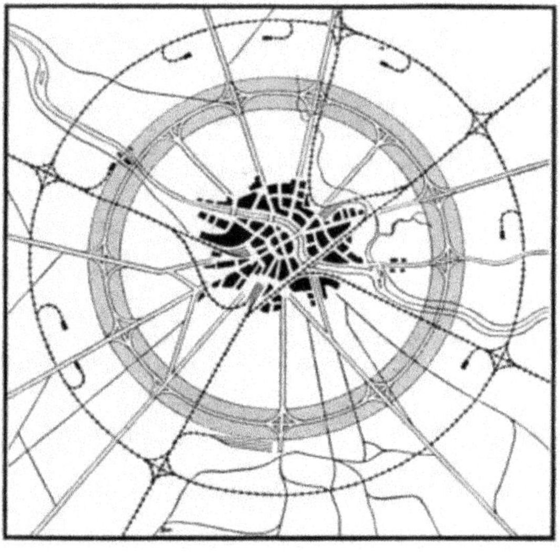

Fig. 2.3 Line drawing of "the Wiener civil defense plan's" proposal for a series of protective "outer belts" that will preserve movement of people and supplies in the wake of a nuclear attack ("How U.S. Cities Can Prepare," 79; image credit: Anthony Sodaro, *LIFE Magazine*).

[75] Deutch, de Santillana, and Wiener, "How U.S. Cities Can Prepare," Norbert Wiener papers, 78–79.

the contrasting comparisons Wiener often deploys when marshalling historical evidence), we witness an early version of the type of speculative design thinking that more recent scholars advocate as necessary if we want to "generate futures that act as catalysts for public debate and discussion about the kinds of futures people really want" and "set in place today factors that will increase the probability of more desirable futures happening." [76] These latter statements appear in Anthony Dunne and Fiona Raby's *Speculative Everything* (2013), a book that opens with a list of design attributes they seek to embrace, many of which resonate with the ambitions that underpin Wiener's writing. Design "in the service of society," "for how the world could be," which adopts a "critical" stance and integrates "humor," "rhetoric," and "ethics"[77]—a more apt description of Wiener's vision for the socially responsible technologist would be hard to find.[78] As they elaborate, "this form of design thrives on imagination and aims to open up new perspectives on what are sometimes called *wicked problems*, to create spaces for discussion and debate about alternative ways of being, and to inspire and encourage people's imaginations to flow freely."[79] The call-out to Horst W.J. Rittel and Melvin M. Webber's notion of a "wicked problem" (originally presented in 1973 and now a mainstay of discourse on complex systems thinking, having arguably become an example of Lakoff and Johnson's "creative metaphor" that has made it into the broader cultural imaginary) likewise amplifies many of the concerns that I have traced in Wiener's mid-century work. As Rittel and Webber frame it, the wicked problem emerges from a recognition that "test[s] for efficiency" are inadequate for measuring "accomplishment" in the context of complex social and policy-focused problems, since

[76] Anthony Dunne and Fiona Raby, *Speculative Everything: Design, Fiction, and Social Dreaming* (MIT Press, 2013), 16.

[77] Dunne and Raby, *Speculative Everything*, 8.

[78] Dunne and Raby also share with Wiener a conviction in the *limits* of speculation as a generative activity. "Predicting or forecasting the future" is something they are "absolutely not interested in," since "when it comes to technology, future predictions have been proven wrong again and again. In our view, it is a *pointless activity*" (Dunne and Raby, *Speculative Everything*, 13. emphasis added). Wiener's "Too Damn Close" piece echoes this tone when he dismissively labels it "*sheer folly* to estimate the social consequences of atomic physics on the strength of what we can knowingly accomplish with the present atomic bombs" ("Too Damn Close," Norbert Wiener papers, 5, emphasis added).

[79] Dunne and Raby, *Speculative Everything*, 13.

efficiency-focused metrics fail to account for more complex "consequences for equity" within a nexus of "open … systemic networks."[80]

Speculation may not solve the problems of technology's inequitable social consequences. Nonetheless, Wiener's work—like the concepts these later authors present—invites us to consider how the ability to think imaginatively about potential future outcomes in advance of taking action is crucial, especially within contexts where "every implemented solution is consequential."[81] He repeatedly offers readers experiences of speculative immersion. These appear not only in the *Life Magazine* article's vision of alternative urban planning projects, but also in places where the subject matter is more technical in scope, as in the final pages of "Newtonian and Bergsonian Time," which present "scientific speculation[s]" on strategies for observing atomic particles in action.[82] Speculation thus emerges as another rich take on analogy-based thinking. This rhetorical tool constitutes a generative approach for readers who seek active ways to engage with contemporary crises but don't wish to sink into bitterness and despondency.

Provisional Conclusions: Rhetorical Toolkits for Humanity

As I noted in my opening statements, the term "disunity" in the title of this essay is drawn from Kline's description of cybernetics. As his and others' work illustrates, cybernetics was notoriously (and, its champions would likely proclaim, *necessarily*) challenging to define as it traversed

[80] Horst W. J. Rittel and Melvin M. Webber, "Dilemmas in a General Theory of Planning," *Policy Sciences* 4, no. 2 (1973): 156. 10.1007/BF01405730. Rittel and Webber make explicit reference to cybernetics in their description of what has come to be accepted as "*how* and *idealized* planning system would function"—namely, as "an on-going, cybernetic process of governance" in which statistical monitoring and feedback loops enable all "errors to be corrected"; however, they posit that this ideal is "unattainable" within the modern world (Rittel and Webber, "Dilemmas in a General Theory of Planning," 159).

[81] Rittel and Webber, "Dilemmas in a General Theory of Planning," 163.

[82] "Newtonian and Bergsonian Time," Norbert Wiener papers, 9. In these passages, Wiener talks about positrons and the fact that you have to watch for "traces of the passing of particles after they have past." Building from the essay's earlier invocation of Carroll's work, he offers explanations through the idea of the rabbits in boxes, noting that new challenges emerge in the subatomic particle experiments, since you need to be able to contain the rabbits, a thing that's not always possible with, say, a "neutron [which] will run through any sort of matter" (10).

traditional boundaries of intellectual inquiry and identified points of connection across "a host of seemingly unrelated subjects."[83] And in many ways, the "disciplinary agility" that I have argued characterizes Wiener's diverse approaches to communicating cybernetics' core analogical premise—whether through humorous and ironic puns, illustrative invocations of poetry, bitingly critical references to Shakespearian tragedy and violence, historical examples and counterpoints, or a speculative imaginary—exemplifies this notion of disunity. As a creative rhetorician, Wiener is not picky about pedigree; he will make use of the full gamut textual strategies available to him to advance his arguments about the necessity of seeing the world through the cybernetics lens.

However, through these readings, I have also come to recognize that a consistent impulse underpins Wiener's rhetorical variety, a fact that belies the "disunity" argument. After all, a repeated refrain throughout my analyses has been that Wiener strives to convey the urgency of his present moment and to convince readers—whether fellow scientists and technologists, colleagues from adjacent or far-flung disciplines, public servants and policymakers, or members of the general public—that they should take seriously and act on that urgency. His work is therefore decidedly ***unified*** in its goal of making an appeal to ethical responsiveness and social responsibility in the face of technology-enabled crisis. While Wiener's efforts at broad-scale culture change might not be classified as widely influential within their immediate social context,[84] I invite us to contemplate how his cross-disciplinary agility might nonetheless assume heightened significance for readers today, embedded as we are in the era of the wicked problem. The ultimate success of his appeals can be debated (as Kline reminds us, the "cybernetics moment" has passed[85]), but his conviction in speaking beyond the purview of the scientific community endures in today's discourses of interdisciplinarity, and the variety of approaches he musters to reach an astonishingly wide range of audiences can be viewed as a type of "rhetorical toolkit" for twenty-first-century readers.

[83] Kline, *The Cybernetics Moment: Or, Why We Call Our Age the Information Age*, 233.

[84] Additional research tracking responses to the specific texts presented in this essay would be needed to make a definitive case for their success in altering public perception or influencing his immediate audiences, but biographical writing by authors such as Conway and Siegelman suggests that Wiener's prickly personality and abrasive approach to communication may have limited the reach and impact of at least some of his ideas. See Conway and Siegelman, *Dark Hero of the Information Age*.

[85] Kline, *The Cybernetics Moment: Or, Why We Call Our Age the Information Age*.

The desire to build "toolkits" that will enable more ethical, responsible, value-based, social justice oriented, or [insert the adjective that best resonates with your goals] approaches to technological innovation is alive and well today,[86] and Wiener would surely have much to say about the extent to which at least some of these efforts are deployed for "ethics washing" rather than meaningful change.[87] But, in the interest of highlighting the humanities-based disciplinary perspective I have foregrounded throughout this piece, I want to close with an invocation of the "toolkit" that resonates strongly with Wiener's work and that is drawn from the field of rhetoric. In his 2024 monograph *Panic Now? Tools for Humanity*, Ira J. Allen offers a take on what he labels the twenty-first-century "polycrisis"—"the heap of crises that will be the end of one world, an accumulating and mutually amplifying collection of deferred costs that cannot be paid."[88] His framing recapitulates in many ways both the concerns/arguments and the rhetoric/stylistics that Wiener articulated three quarters of a century earlier. Wiener's alliteratively articulated focus on automation and atomic warfare as the root of the crises he perceived as most in need of attention at mid-century becomes transposed, in Allen's argument, to today's trifecta of "carbon, capitalism, and colonialism (or CaCaCo)."[89] Wiener's condemnation of political and military leaders who fail to act appropriately finds an echo in Allen's observations that "relative to the actual scale of the [climate] crisis, even the 'largest climate bill ever' is fundamentally unserious."[90] And Wiener's unapologetic insistence on a vocabulary of catastrophe is a mainstay of Allen's writing. "So, why panic now?" he queries; "Because our world is ending."[91]

But like in Wiener's writing—with its willingness to show us glimmers of humor and hope amid the crisis—Allen's "Preface" also clarifies that the instruction to panic is meant as a generative invitation. After all, he posits

[86] For one inventory of 81 resources that provide such toolkits/frameworks/etc., see the "Resources Database" assembled by the Critical by Design research team (of which I am part) at the University of Waterloo: *Resources – CRITICAL BY DESIGN*, n.d., accessed November 13, 2025, https://criticalbydesign.ca/resources/.

[87] See Karen Hao, "In 2020, Let's Stop AI Ethics-Washing and Actually Do Something," *MIT Technology Review*, December 27, 2019, https://www.technologyreview.com/2019/12/27/57/ai-ethics-washing-time-to-act/.

[88] Ira J. Allen, *Panic Now? Tools for Humanizing* (University of Tennessee Press, 2024), ix.

[89] Allen, *Panic Now? Tools for Humanizing*, ix.

[90] Allen, *Panic Now? Tools for Humanizing*, 7.

[91] Allen, *Panic Now? Tools for Humanizing*, ix.

with care, "panicking now lays the only groundwork there could be for genuinely better next worlds. May we build wisely among the ruins of this one."[92] In this provocative toggling between seemingly incompatible stances (the world is ending/let us build better worlds), we locate the heart of Wiener's rhetorical ethos. Wiener's writing demonstrates how compelling arguments about modernity, technology, responsibility, and agency emerge from a precarious balance between seemingly opposed perspectives—from actively *making* analogy-based or metaphorically articulated comparisons while arguing for their *already-existing* status; by drawing on historical genealogy to illustrate both the *continuity* of uninterrupted lineage and the *contrast* of socio-political rupture; through acknowledging *likely* disastrous futures while crafting the *possibility* of more hopeful alternatives through speculation. Within the "contestations of the [present]," he teaches us, these intricate rhetorical dances may be some of the most important "paths [we might] take."

Bibliography

Adamson, Greg, Ronald R. Kline, Katina Michael, and M. G. Michael. Wiener's Cybernetics Legacy and the Growing Need for the Interdisciplinary Approach [Scanning Our Past]. *Proceedings of the IEEE* 103, no. 11 (2015): 2208–14. https://doi.org/10.1109/JPROC.2015.2483958.

Allen, Ira J. *Panic Now? Tools for Humanizing*. University of Tennessee Press, 2024.

Conway, Flo, and Jim Siegelman. *Dark Hero of the Information Age*. Basic Books, 2009. https://www.hachettebookgroup.com/titles/flo-conway/dark-hero-of-the-information-age/9780786735617/?lens=basic-books

Dunne, Anthony, and Fiona Raby. *Speculative Everything: Design, Fiction, and Social Dreaming*. MIT Press, 2013.

Guillory, John. *On Close Reading*. With Scott Newstok. University of Chicago Press, 2025. https://press.uchicago.edu/ucp/books/book/chicago/O/bo239363263.html

Hao, Karen. "In 2020, Let's Stop AI Ethics-Washing and Actually Do Something." *MIT Technology Review*, December 27, 2019. https://www.technologyreview.com/2019/12/27/57/ai-ethics-washing-time-to-act/

Hills, David. "Metaphor." In *The Stanford Encyclopedia of Philosophy*, Fall 2024 Edition. August 19, 2011. https://plato.stanford.edu/archives/fall2024/entries/metaphor/

[92] Allen, *Panic Now? Tools for Humanizing*, xi.

Kline, Ronald R. *The Cybernetics Moment: Or, Why We Call Our Age the Information Age*. Johns Hopkins University Press, 2015b.

Lackoff, George, and Mark Johnson. *Metaphors We Live By*. The University of Chicago Press, 1980.

Lafontaine, Céline. The Cybernetic Matrix of 'French Theory'. *Theory, Culture & Society* 24, no. 5 (2007): 27–46. https://doi.org/10.1177/0263276407084637.

Love, Heather A. *Cybernetic Aesthetics: Modernist Networks of Information and Data*. Cambridge University Press, 2023. https://doi.org/10.1017/9781009387446.

Love, Heather A. "The Cybernetics Moment: Or Why We Call Our Age the Information Age by Ronald R. Klein (Review)." Reviews. *Configurations* (Baltimore, United States) 25, no. 4 (2017): 557–59. https://doi.org/10.1353/con.2017.0035.

Love, Heather A., Greg Adamson, Mallory James, et al. "The Future of Work in the Age of Automation: Proceedings of a Workshop on Norbert Wiener's 21st Century Legacy." *IEEE Transactions on Technology and Society*, 2024, 1–23. https://doi.org/10.1109/TTS.2024.3476041.

Mills, Mara. On Disability and Cybernetics: Helen Keller, Norbert Wiener, and the Hearing Glove. *Differences* 22, 2–3 (2011): 74–111. https://doi.org/10.1215/10407391-1428852.

Newstok, Scott. 2025. A Century of 'Close Reading.' *Public Humanities* 1: e90. https://doi.org/10.1017/pub.2025.13.

Pias, Claus, ed. *Cybernetics: The Macy Conferences 1946–1953. The Complete Transactions*. DIAPHANES, 2016. https://press.uchicago.edu/ucp/books/book/distributed/C/bo23348570.html

Pickering, Andrew. *The Cybernetic Brain: Sketches of Another Future*. University of Chicago Press, 2011. https://press.uchicago.edu/ucp/books/book/chicago/C/bo8169881.html

Resources – CRITICAL BY DESIGN. n.d. Accessed 13 November 2025. https://criticalbydesign.ca/resources/

Rittel, Horst W. J., and Melvin M. Webber. "Dilemmas in a General Theory of Planning." *Policy Sciences* 4, no. 2 (1973): 155–69. https://doi.org/10.1007/BF01405730.

Wiener, Norbert. *Cybernetics: Or Control and Communication in the Animal and Machine*. Wiley, 1948.

Wiener, Norbert. *The Human Use of Human Beings: Cybernetics and Society*. Houghton Mifflin, 1950.

Wiener, Norbert. n.d. Norbert Wiener papers. Massachusetts Institute of Technology. Libraries. Department of Distinctive Collections. Cambridge, MA.

CHAPTER 3

The Strange Story of Automotive Collabs and Oil Embargoes

Rayvon Fouché

INTRODUCTION

Writing an article in honor of someone is an uncomfortably challenging thing to do. Writing an article in honor of my dissertation advisor's collective scholarly output is a fundamentally terrifying endeavor. As I write these words, I keep telling myself that I am up to the task—I hope. I assume I, like many others, probably scoured books, journals, articles, and all things googleable to find examples, references, and guidance of how to execute this task well. For the best counsel, I tip my cap to David Schleicher, whose 2013 *Tulsa Law Review* article, "From Here All-the-Way-Down, or How to Write a Festschrift Piece," was the most informative essay I read

R. Fouché (✉)
Northwestern University, Evanston, IL, USA
e-mail: fouche@northwestern.edu

J. L. Lieberman, A. Mohsin (eds.), *Technological Humanities*,
https://doi.org/10.1007/978-3-032-24741-4_3

and gave me the courage to dive in with both feet.[1] This essay will loosely follow Schleicher's three-piece framework of "[1] fawning praise…[2] find a problem or a hole somewhere in the honoree's body of scholarship preferably a minor one…[3] argue that the problem identified in Section II, could be improved by reference and discussion of themes that appear in your own work."[2]

The praise begins with Ron's (I call him Ron, because I have rarely ever referred to him in any other way) ability to understand the way everyday people use, understand, and experience technology.[3] Though a significant portion of the history of technology focuses on large-scale systems, networks, and infrastructures (think Thomas Hughes' *Networks of Power*), I think Ron and I share the interest in smaller, more digestible slices of the technological world and focus on people and the technological objects that inhabit their lives. One of the most important things I take from Ron is to study and embrace the way folks, not seen as the technological powerbrokers, creatively integrate technologies into their lives based on their own specific needs, wants, and desires. From Ron's work, I learned that the technology is important, but the people are more important. In my mind, the rephrasing and adjustment of the familiar history of technology phrase "follow the technology," to the wordier, but for me, more useful phrase "follow the technology, but look hard at what those marginalized (race/gender/class/etc.) are doing with the technology!" is the key insight. It is about understanding how and why people engineer, design, make, build, consume, and use technology for a host of fascinating and convoluted reasons. Yes, the larger world can scuttle the best-laid plans, but people are resilient and can be very motivated to use technologies in ways that undermine or downright transgress the dominant meaning and use for a technological device or system.[4]

---

[1] David Schleicher, "From Here All-the-Way-Down, or How to Write a Festschrift Piece," *Tulsa Law Review* 48, no. 3 (2013): 401–25.

[2] Schleicher, 401–402.

[3] Kline, *Consumers in the Country*.

[4] Rayvon Fouché, *Black Inventors in the Age of Segregation* (Johns Hopkins University Press, 2005), 10.56021/9780801873195; Rayvon Fouché, *Game Changer: The Technoscientific Revolution in Sports* (Johns Hopkins University Press, 2017); Rayvon Fouché, "The Wretched of the Gulf: Racism, Technological Dramas, and Black Politics of Technology," *The Black Scholar* 36, no. 4 (2006): 7–12; Rayvon Fouché, "Say It Loud, I'm Black and I'm Proud: African Americans, American Artifactual Culture, and Black Vernacular Technological Creativity," *American Quarterly* 58, no. 3 (2006): 639–61.

My point of entry for step two—finding a small hole section—is the automobile.[5] I have a long and loving relationship with cars. I thought very hard about using the deeply emotional word *love* to describe my connection to cars, but it's appropriate because my fascination with and interest in cars runs deep. From temporarily pausing a cross-country trip when a blizzard shut down highway I-90 in Idaho to spending a day with my then 17-year-old son installing way-too-low lowering springs on his carefully modded Subaru STI, some of my long-lasting childhood and adult memories connect to experiences of riding in, driving, fixing, and customizing cars.

When I entered graduate school in the early 1990s, it was very clear to me that the academic world in general did not share my deep love, affection, and appreciation for the automobile. To this day, nothing can get my colleagues' eyes to glaze over more quickly than talking about automobiles in a non-functional transportation way, so I generally keep quiet about my automotive adoration.[6] Needless to say, when I learned that Ron and Trevor Pinch planned to publish an article that centered on automobiles, I was *extremely* excited to read it. That article, "Users as Agents of Technological Change: The Social Construction of the Automobile in the Rural United States," was and still is one of the few STS articles that makes the automobile the central character of the story.

Well before the recent reemergence of electric automobiles and electrification euphoria, and the associated conversations about renewable energy, autonomous vehicles, and rare-earth elements, Ron's work endeavors to study the automobile and "...how actions taken by users resulted in changes to both the interpretation and design of an artifact considered to be relatively stable."[7] As much as I love this article, I also feel it did not go far enough in conceptualizing the connections between the automobile and society to explore the strange and fascinating ways automobiles are a productive site to study the ways human creativity, technological dreams, and economic desires merge. Of course, this is not the intent of the article, which made an important theoretical argument about the social construc-

[5] In this essay, I use the more colloquial term *car* interchangeably with automobile.

[6] Though, I do share this appreciation and affection with one colleague, Lisa Nakamura. It is always such a pleasing moment to me when Lisa reaches out to me to talk about cars. But generally, she's the only one.

[7] Ronald Kline and Trevor Pinch, "Users as Agents of Technological Change: The Social Construction of the Automobile in the Rural United States," *Technology and Culture* 37, no. 4 (1996): 763–95.

tion of technology. Nevertheless, to take Ron and me to the third component of Schleicher's three-step process, I want to add a different texture to the automobile as a technology and its situatedness within the sociopolitical milieu of the late 1960s and early 1970s.

Drawing inspiration from Ron's work, this essay aims to reconsider how the 1973 oil embargo and increased competition from Japanese manufacturers forced American automobile companies to rethink their product lines and how they would sell cars to American consumers. A narrow economic history could explain this situation rather quickly, but, arguably, flatly.[8] For instance, that history could explain this moment through gasoline—how a significant increase in its price required American automobile manufacturers to move away from large, heavy, and high-performing cars as the centerpieces of their sales strategy. A basic gasoline-focused narrative about how automobile manufacturers reshaped their offerings in order to continue moving automobiles to consumers leaves much to be desired. Embracing what I learned from Ron to think broadly about not only how people use automobiles but also how they integrate them into their lives as social, cultural, political, and even aesthetic artifacts demands a fuller-textured reading of this technological artifact at this specific moment in the history of the United States.

The bottom line for automobile manufacturers, regardless of national origin, is to sell cars, but they also are playing a longer game that invests in the manipulation of customer loyalty, consumer desires, performance cravings, and technological fantasies.[9] The early 1970s restructuring of the automotive marketplace demanded American manufacturers reconsider how to sell cars that did not fulfill historic consumer expectations. That is, how do you get people to pay up for something that is less than, in nearly every possible way, its predecessor? In this environment, American cars needed a new hook. This was a tall order for the industry, which, beginning at the end of World War II, fed American consumers a steady diet of dynamic designs laced with expanding horsepower.[10] Addicting consumers to this seductive automotive elixir was the basis of the industry's

[8] Lorraine Eden and Maureen Appel Molot, "Made in America? The US Auto Industry, 1955-95," *The International Executive* 38, no. 4 (1996): 501–41, 10.1002/tie.5060380406.

[9] Wolfgang Sachs, *For Love of the Automobile Looking Back into the History of Our Desire*, trans. Don Reneau (University of California Press, 1992).

[10] James J. Flink, "Three Stages of American Automobile Consciousness," *American Quarterly* 24, no. 4 (1972): 451–73, 10.2307/2711684.

profitability.[11] This approach succeeded for decades. But as environmental considerations and oil geopolitics altered the landscape of automobile use and consumption, it was going to be a big ask to get consumers to cut back on thrilling design aesthetics and performance.

To address this reality during the 1970s oil crisis, automobile manufacturers refashioned their automobiles to be less artifacts of American exceptionalism and more objects of consumer amplification. American manufacturers created a new sense of desire that did not necessarily align with the automotive industry's history and tradition. Manufacturers believed one potential avenue was to create what James H. Kaster would name "couture cars," or cross-branded and collaboratively designed limited-edition automobiles.[12] This idea has a long history in the automotive industry. The craft of designing and building custom exteriors for automobiles that differentiated them from others, or what is known as coachbuilding, was a key aspect of the early automotive business. By the 1920s it was such a thriving cottage industry that General Motors president Alfred P. Sloan deemed it wise to create the "Art and Color Section" of General Motors (GM) to capture and capitalize on the coachbuilding creative energy that siphoned off profits from GM. Sloan appointed Harley Earl, who became one of the most important automotive designers of the twentieth century, as its first director.[13]

The industry has always thought deeply about style and design, but it would take a geopolitical crisis to get American automobile manufacturers to build new non-automotive partnerships and leverage the name recognition of fashion houses and brands—such as Cartier, Givenchy, Gucci, Levi's, and Bill Blass—to make certain versions of their cars more attractive to consumers in challenging economic environments.[14] Though short-lived, these large-scale cross-branding efforts would lay an important foundation for future technoscientific/fashion collaborations.

[11] Ann F. Friedlaender et al., "Costs, Technology, and Productivity in the U.S. Automobile Industry," *The Bell Journal of Economics* 14, no. 1 (1983): 1–20, 10.2307/3003534.

[12] James Kaster, *American Automotive Design Trends / The Couture Car: High Style for High Society* (Lulu.com, 2011).

[13] Jonathan A. Stein, ed., *The Art and Colour of General Motors* (Coachbuilt Press, 2008).

[14] Ricard Bennett, "The 1979 Lincoln Continental Mark V Was a Designer's Delight," May 22, 2019, https://www.hagerty.com/media/opinion/1979-lincoln-continental-mark-v-designers-delight/.

Embargoing Oil

To understand how the 1970s oil crisis motivated the auto industry to collaborate with fashion brands, it is valuable to understand the historical context that produced the oil crisis in the first place. Of the multiple events that precipitated the difficulties, the essay will focus on three specific catalysts: (1) late 1960s and early 1970s American politics, (2) global macroeconomic conditions, and (3) pressures felt by the American auto industry. All three contributed to the 1973 oil embargo and reshaped the American auto industry to this very day.

We can trace the political tensions that led to the 1973 oil embargo to the Israeli Declaration of Independence in 1948. Since that moment, conflict and strife between Arabs and Israelis have been a consistent component of Middle Eastern life, resulting in war and countless lives lost.[15] In many of these crises, the Suez Canal was a central site of conflict. The impetus for the 1956 Suez Crisis was Egyptian nationalization of the Suez Canal and the blocking of Israel's southern port of Eilat. The conflict would close the canal for a substantial part of 1956 and 1957. In 1967, during the Six-Day War, Israel invaded the Egyptian Sinai Peninsula, which caused Egypt to close the Suez Canal for eight years. The canal did not reopen until 1975 and the end of the Yom Kippur War.[16]

It is important to remember that oil was and still is a global commodity that has been a political fulcrum during most of the twentieth century, and the Middle East is just one of many locations where the struggle to control oil is alive and well.[17] When the United Kingdom, France and Israel attacked Egypt in 1956, Arab oil-producing countries did not sit idly by. Syria destabilized the flow of oil to Europe when it attacked the Trans-Arabian and the Iraq–Baniyas pipelines.[18] By the early 1970s, oil would become the central actor in geopolitical warfare and the OAPEC (Organization of Arab Petroleum Exporting Countries) would lead the movement.[19]

[15] Including 1956 Suez Crisis, 1967 Six-Day War, 1973 Yom Kippur War, 1982 Lebanon War, and 2006 Lebanon War.

[16] Fiona Venn, *The Oil Crisis* (Routledge, 2002).

[17] During the drafting of this essay, the current United States administration spuriously staked its claim on Venezuelan oil.

[18] Rafael Kandiyoti, *Pipelines: Flowing Oil and Crude Politics* (I.B.Tauris, 2008).

[19] OAPEC should not be confused with the OPEC (Organization of the Petroleum Exporting Countries).

In 1968, as global demand increased, Kuwait, Libya, and Saudi Arabia founded OAPEC as an intergovernmental organization charged with managing the region's concerns about the weaponization of oil.[20] This organization initially mobilized Arab countries to stabilize the export market by curtailing the use of embargoes or other tactics that could undermine oil as a driver of wealth. It also promoted the political and economic interests of its Arab oil-exporting members, which is why Egypt and Algeria, countries seen as too radical, were not offered membership. Iraq found the organization too passive and declined to join. By 1972, the changing landscape of the Middle East led OAPEC to add Algeria, Iraq, Syria and Egypt and take a more radical stance supporting a collective Middle Eastern political agenda that embraced using oil as a geopolitical tool, which was put into action during the Arab-Israeli conflict.

The organization became a global political and economic force in 1973 when, during the Yom Kippur War, Egypt and member countries collectively agreed to decrease oil shipments to countries supporting Israel. OAPEC agreed to shrink oil production by 5% monthly "until the Israeli forces…completely evacuated…all the Arab territories occupied in the June 1967 war."[21] The embargo led to substantial increases in oil prices across the globe and forced countries dependent on importing oil from the Middle East to scramble to fulfill their growing demand for petroleum products. With the United States and many of its Western allies supporting Israel, the move significantly ruffled international oil markets. The embargo lasted approximately five months before OAPEC rescinded it in March 1974 after negotiations at the Washington Energy Conference in February 1974.[22] This collective action would empower these countries and reshape the political and economic landscape for the remaining decades of the twentieth century. For most oil-exporting countries, the interaction revealed a newfound power in limiting oil to countries whose citizens had become used to having ready access to its products, most significantly, gasoline. For the United States, a country that marked its global ascendancy through the construction of a national highway system

[20] George Tomeh, "Oapec: Its Growing Role in Arab and World Affairs," *The Journal of Energy and Development* 3, no. 1 (1977): 26–36.

[21] A. F. Alhajji, "Three Decades After the Oil Embargo: Was 1973 Unique?," *The Journal of Energy and Development* 30, no. 2 (2005): 223–37.

[22] Rüdiger Graf, "Making Use of the 'Oil Weapon': Western Industrialized Countries and Arab Petropolitics in 1973–1974," *Diplomatic History* 36, no. 1 (2012): 202.

that required the use of automobiles, the obsession with the car would not only be a signifier of strength, but also a geopolitical weakness.

Redefining the American Car

The 1973 oil crisis marked a substantive turning point in the trajectory of the American automobile industry. Since the end of the Second World War, Detroit's "Big Three"—General Motors, Ford, and Chrysler—dominated the global automotive marketplace with large, flashy, powerful vehicles that symbolized postwar affluence, American exceptionalism, technological superiority, and leading-edge design aesthetics. The sudden fuel price increases of 1973, coupled with the rising prominence of Japanese automakers such as Datsun, Honda, and Toyota, revealed profound vulnerabilities for the American automotive industry's long-term sustainability.[23] This was a watershed moment. The crisis did not simply disrupt the industry's business model; it also reshaped the aesthetics, design philosophies, and cultural meanings of American automobiles. This transition demanded that the automotive industry rethink the tensions between automotive transport and automotive joy.

The disjuncture between automobile ownership as a pleasurable leisure activity and automobile use as functional transport defines the disquietude the oil crisis presented. The automotive industry worked hard to unite Americanized ideals of independence, freedom, and mobility with emotional and joyful motoring behind the wheel of a large, powerful vehicle.[24] From the appropriative westward expansion across the continent to the development of the interstate highway system in the twentieth century, the perceived freedom of movement across vast spaces shaped a national imagination.[25] This expansion made possible by technological force aligns well with the automobile and the sentiment that driving a powerful vehicle allows an individual to feel an autonomous control of motion and direction, reinforcing a cultural narrative that values self-determination and

[23] Akira Kawahara, *The Origin of Competitive Strength: Fifty Years of the Auto Industry in Japan and the U.S.*, reprint of the original (Springer, 2012).

[24] Mimi Sheller, "Automotive Emotions: Feeling the Car," *Theory, Culture & Society* 21, nos. 4–5 (2004): 221–42, 10.1177/0263276404046068.

[25] Mark H. Rose, *Interstate: Express Highway Politics 1939-1989* (University of Tennessee Press, 1990).

personal choice.[26] Advertising effectively played a key role in conjuring these sentiments and desires by fusing them with the American understanding of automobiles. Beyond the American cultural automotive imaginaries, there is also joy in the physical sensations experienced while driving. The sound of an internal combustion engine, its powerful acceleration, and the feel of the road can be exhilarating. These sensory pleasures merge technological power with mechanical mastery. For some drivers, this experience affirms a belief in progress, abundance, and control of the natural world. But the oil crisis abruptly frayed and eventually broke these connections for many automobile consumers.

The pressure put on markets reliant on oil by the 1973 oil crisis was profound. The Organization of Petroleum Exporting Countries (OPEC)—again different from the OAPEC—1973 oil embargo caused crude oil prices to surge from $2.75 per barrel in early 1973 to over $11 in 1974.[27] American consumers, who had long favored plush-riding sedans, voluminous station wagons, and powerful V-8 muscle cars, suddenly confronted long lines and unprecedented price volatility precipitated by gasoline shortages.[28] The 1970s energy crisis directly caused a significant collapse in demand for most of the American manufacturers' flagship models, which, not surprisingly, were their most profitable, and exposed the inflexibility of an industry structured around the belief that the demand for size, power, and petroleum would never end.

Designing a New Future

As the clean lines of mid-century modern architecture emerged in the middle part of the twentieth century, automotive design still clung tightly to older design cues. Designers such as Harley Earl endeavored to reshape and redefine what a modern American automobile could be, but it was difficult to get change-averse corporations and their well-trained consumers to buy into mid-century modern design aesthetics and disavow expansive chrome grilles, oversized tailfins, and high-displacement engines that

[26] James A. Dunn Jr, *Driving Forces: The Automobile, Its Enemies, and the Politics of Mobility* (Brookings Institution Press, 1998).

[27] Jack L. Hervey, *The 1973 Oil Crisis: One Generation and Counting*, Chicago Fed Letter 86 (1994), https://www.chicagofed.org/publications/chicago-fed-letter/1994/october-86.

[28] Meg Jacobs, *Panic at the Pump: The Energy Crisis and the Transformation of American Politics in the 1970s* (Hill and Wang, 2016), 111.

symbolized an Americanized prosperity.[29] Tastes would change throughout the 1960s and by the early 1970s, the familiar design language appeared impractical, irresponsible, and, for many, just no longer desirable. Fuel efficiency became not only an economic necessity but also a cultural value that forced manufacturers to reconsider the very proportions and styling of their products.[30] In terms of design, the crisis initiated a break with the longer, lower, and wider aesthetics that defined the American automotive marketplace since the 1950s. Japanese automakers capitalized on these global economic shifts and designed cars that fit the new environmentalist cultural sensibilities emerging in the 1960s, and the first Earth Day, commemorated on April 22, 1970, proved the environmental movement had matured.[31]

Japanese manufacturers entered the United States market from the late 1950s to early 1960s with middling success. But in 1968, with the introduction of the Toyota Corolla and the Datsun 510 and the changing cultural landscape, Japanese automakers began to make some headway in the American market. The Corolla and the Honda Civic, introduced to the United States in 1973, would come to epitomize automotive efficiency, simplicity, and reliability for decades. Their compact size and lightweight made them capable of achieving 30 to 40 miles per gallon (mpg) where their American counterparts struggled to get anywhere close to that level of fuel efficiency. They also tapped into an emerging American zeitgeist that desired the aesthetic rationality and restraint Japanese cars represented, which was decidedly different from the cultural tumult of the 1960s. Their uncluttered lines, smaller footprints, and minimal ornamentation contrasted sharply with an American design language that championed an aesthetic of excess. Japanese cars expressed a less is more ideal and encouraged a design language emphasizing aesthetic grace, utility, and balance that removed extraneous elements and embraced a minimalism in form and function. Quality became a technological design feature. Japanese cars, criticized and castigated as subpar in the post-World War II years, were, by the late 1970s, consistently ranked above American cars in

[29] Jeffrey I. Godshall et al., "The Automobile - Unwanted Technology: Part II the Middle Years (1930–1960)," *SAE Transactions* 100, no. 6 (1991): 1353–58, 10.4271/910903.

[30] John R. Pierce, "The Fuel Consumption of Automobiles," *Scientific American* 232, no. 1 (1975): 34–47.

[31] Adam Rome, *The Genius of Earth Day: How a 1970 Teach-In Unexpectedly Made the First Green Generation* (Hill and Wang, 2013).

durability and fit-and-finish.[32] As a result, the American car buying public began to associate Japanese automotive manufacturing and design—precise panel gaps, understated interiors, efficient layouts—with a useful and functional modernism. This aesthetic success reconfigured consumer expectations and the association of Japanese-rooted lean design with quality and reliability.

The 1973 oil crisis and subsequent Japanese competition destabilized the American auto industry's reliance on large, energy-inefficient cars. Japanese automakers demonstrated that, during an energy crisis, efficiency could resonate deeply with consumers. Detroit's eventual embrace of downsizing marked a paradigm shift in American automobile aesthetics. What emerged was a new design philosophy that fused consumer pragmatism with industrial necessity. The oil crisis stands as not only an energy shock, but also an aesthetic revolution—one that permanently redefined what an American car looked like and meant in a more interrelated globalizing world.

As much as cultural tastes and sensibilities motivated automotive consumption and the global realities of oil reshaped the automotive landscape, federal policy played an important role in refashioning American automotive manufacturing. Specifically, the 1975 Energy Policy and Conservation Act introduced the Corporate Average Fuel Economy (CAFE) standards, which required automakers to reach 18 mpg by 1978 and 27.5 mpg by 1985.[33] For most American automobile manufacturers, this was a tall order. To comply, American manufacturers launched a new era of downsizing everything; small and light became a central design principle and a corporate imperative. Iconic designs and beloved shapes no longer held a special place within most manufacturers' product lineups. Companies attempted to redesign their full-size offerings—like the Chevrolet Impala, Ford LTD, and Chrysler New Yorker—to deliver smaller versions in every possible way (shorter wheelbases, lighter bodies, and smaller and less powerful engines with better fuel efficiency). They would hit production lines in the early 1980s. This shift altered American car aesthetics profoundly. For many American consumers, this left much

[32] Brad M. Barber and Masako N. Darrough, "Product Reliability and Firm Value: The Experience of American and Japanese Automakers, 1973-1992," *Journal of Political Economy* 104, no. 5 (1996): 1084–99.

[33] Laura Hall, "The Evolution of CAFE Standards: Fuel Economy Regulation Enters Its Second Act," *Transportation Law Journal* 39, no. 1 (2011): 1–30.

to be desired. Instead of elongated hoods and regal chrome trim, design aesthetics turned toward weight reduction and efficiency. Interiors also became less ornate, reflecting perceived consumer preferences and needs for practicality over luxury.

From the start, Detroit's manufacturers struggled to compete against Japanese imports. It was hard for the American automakers to reconfigure their expansive production lines. It was even harder for these hierarchical organizations to make the conceptual and cognitive adjustments necessary to embrace an automotive aesthetic seen as un-American. They introduced compact models to compete with imports, Ford's Pinto and later the Escort, GM's Chevrolet Vega and J-body compacts, and Chrysler's Dodge Omni/Plymouth Horizon, but the speed at which these cars entered the market did not help their cause. Quality issues and safety flaws plagued these rushed-to-market vehicles and undercut their desirability and competitiveness.[34] It would take nearly another decade before American manufacturers would get it right with Ford's release of the European-designed Taurus in 1986. Its rounded, aerodynamic silhouette departed from the American tradition of boxy bodies. The Taurus demonstrated that American manufacturers could meet efficiency demands with functional elegance, but it was over a decade too late.

Collaborative Motoring

By the mid-1970s, a global recession enveloped the world and American consumers were looking for economic relief.[35] American automakers suddenly found themselves beset with soaring fuel prices, environmental regulation, and the surging popularity of gas-saving, lightweight, nimble, and "cuter" cars from Japan amid an economic downturn. The traditional sales formula of large engines, elaborate styling, and sheer physical size and the racing mantra of "win on Sunday, sell on Monday," no longer guaranteed consumer brand and purchasing loyalty.[36] In response,

[34] National Research Council, *The Competitive Status of the U.S. Auto Industry: A Study of the Influences of Technology in Determining International Industrial Competitive Advantage* (National Academies Press, 1982), 10.17226/291.

[35] María Vázquez-Fariñas, "Major Economic Recessions in the Last Quarter of the 20th Century: The Oil Crisis (1973–1980)," in *The Age of Global Economic Crises*, ed. Juan Manuel Matés-Barco and María Vázquez-Fariñas (Routledge, 2023).

[36] Tom Jensen, "Win on Sunday, Sell on Monday," NASCAR Hall of Fame, June 23, 2023, https://www.nascarhall.com/blog/win-on-sunday-sell-on-monday.

American automakers experimented with new ways of motivating and selling desire. One of the most intriguing strategies was to develop what would eventually be called "couture cars." These were limited-edition vehicles created in partnership with internationally recognized fashion designers and brands.

These collaborations were not expressions of American technological acumen or engineering innovation as much as they were experimental forays in modern cultural branding. By borrowing the aura of haute couture and channeling this energy into customized editions of American automobiles, American automakers amplified cars into lifestyle accessories and objects of conspicuous consumption that symbolized taste, style, and refinement. If American automobile manufacturers were unable to compete with Japanese technology and engineering, they believed, or at the very least hoped, they could "out class" Japanese import cars by linking them with prestigious American or European fashion houses and brands through collaborative marketing.

Currently, "collabs" are a central element of nearly every company's marketing strategy, but this maneuver has a long and rich history.[37] Globally, the idea of "co-branding" an object, device, or technology with a cooler, hipper, or more desirable company, brand, or individual is firmly rooted in these American automotive collaborations. Couture cars of the 1970s and 1980s responded to economic upheaval and cultural change. The primary goal was to delimit the slipping desirability of American automobiles and regain market dominance by merging fashion's prestige with Detroit's manufacturing legacy.

The term couture cars is something of a misnomer. Though it purposely alludes to early twentieth-century vehicles handmade by skilled artisans working for companies such as Auburn, Cord, or Duesenberg, that is where the similarities end.[38] The 1970s-era couture cars were not handcrafted masterpieces in the artisanal sense of couture automobiles or fashion. Most of the 1970s-era couture cars were mechanically indistinguishable from their mass-produced siblings. What set them apart were the details, such as custom paint schemes and plush interior fabrics, but most importantly, designer logos referencing prestigious or familiar fashion

[37] Jagdish N. Sheth and Atul Parvatiyar, "The Evolution of Relationship Marketing," *International Business Review* 4, no. 4 (1995): 397–418, 10.1016/0969-5931(95)00018-6.

[38] Matthew C. Sonfield, "Custom Automotive Coachbuilding in the United States, 1900-1940," *Design Issues* 12, no. 2 (1996): 47–60, 10.2307/1511713.

brands. The message the cars sent was explicit: purchasing one of these vehicles was to buy not merely a mode of transportation, but to inhabit and travel within a recognized and esteemed branded lifestyle. This transformation was an important break for automobiles circumscribed as a mode of transportation rooted within engineering design and technoscientific rationalities. As American manufacturers lost ground on technology and efficiency, companies pivoted to build upon old traditions of cars as status symbols and to recontextualize automobiles as desirable popular cultural artifacts. In a moment where advanced technology, physical stature, and performance were no longer strong selling points, American Motor Company (AMC) exemplifies the automotive arc of a company that needed to go all in on building a brand collaboration highway to prosperity.[39]

AMC, founded on May 1, 1954, resulted from the merger of Nash-Kelvinator Corporation and Hudson Motor Car Company. To get a sense of the size and scale of the company, consider that it was the largest merger in American corporate history at the time.[40] What makes AMC interesting is that the company concentrated on building and selling smaller vehicles from the beginning. Under the leadership of George Romney (the father of former presidential candidate Mitt Romney), AMC had great success in the late 1950s, controlling costs by manufacturing small and economical cars. Even into the 1960s, AMC led the American automotive industry in developing small cars and a host of innovative technologies. AMC lost its way after Romney resigned to run for Governor of Michigan in 1962. His successor, Roy Abernethy, directed AMC to move away from its small car mission and build larger cars to compete with Ford, General Motors, and Chrysler. This went horribly wrong for AMC, and Abernethy "retired" in 1966 under board pressure. Roy Chapin Jr. succeeded Abernethy and returned the company to its small car roots, in part to stabilize the

[39] This was not the first collaboration of the era by any means. Ford and American 24 hours of Lemans winner, Carrol Shelby, cross-branded the GT 350 and the GT500 Shelby Mustangs from 1965–1970. These cars are unique in their own rights and were built in very limited quantities (in the range of 7000) which is approximately half the number of Lincoln Continental Mark IV Bill Blass edition manufactured in 1976 and production increase to nearly 10,000 in 1979. Couture cars, though special, were intended to be manufactured and sold in volume.

[40] Chen Liu et al., "When Transformation Fails: Twelve Case Studies in the American Automobile Industry," *Journal of Enterprise Transformation* 5, no. 2 (2015): 71–112, 10.1080/19488289.2015.1019654.

company by returning to historic strengths, but it was also a plan aimed at refocusing its marketing to sell a younger consumer on AMC. Chapin's leadership coincided with the hiring of the advertising firm Wells, Rich, and Green, headed by Mary Wells Lawrence, who spearheaded AMC's new sales and media strategy.[41]

Through the combined effort of AMC's executive leadership and Wells, Rich, and Greene, AMC would launch a series of collaborations with European fashion brands Pierre Cardin and Gucci. Primarily confined to vivid interior color palettes and sumptuous fabrics, AMC introduced the Gucci edition Hornet Sportabout wagon and the Pierre Cardin Javelin in 1972 and 1973 respectively. AMC championed these collaborative efforts, and in 1973 empowered Vice President of Marketing, William McNealy, to create an interior for the Gremlin that they hoped would have broader consumer appeal. Together with AMC's head of interior design, Vince Geraci, they launched a collaboration with Levi's. This was not without its own challenges. Levi's wanted the interior to "have the design work and the sewing style and the pleating and everything that goes" into the look and feel of a pair of jeans, including "the buttons, the orange stitching, and the little tag that says 'Levi's.'"[42] But Levi's denim was not feasible for a host of reasons ranging from the general long-term durability of cotton denim in a car interior to the fact that denim would not pass federal flame-retardant regulations. They would eventually overcome these issues by working with Milliken & Company, one of the major suppliers of automotive interior fabrics. AMC would even use this interior to produce a series of Levi's Jeeps.

The potential appeal of couture cars becomes clearer when situated within the economic and cultural backdrop of the 1970s. The 1973 oil embargo exposed the vulnerability of America's fuel-hungry personal automobile fleet, leading consumers to reconsider the practicality of large sedans in light of increasingly robust competition from Japanese manufacturers. The American automotive industry faced not only declining sales

[41] "U.S. Business: Irreverence at American," *TIME*, September 22, 1967, https://web.archive.org/web/20110123111601/http:/www.time.com/time/magazine/article/0,9171,837337,00.html. Mary Wells Lawrence firm was one of the most important advertising agencies of the 1970s and it was responsible for some of the most iconic TV and print ads such as "plop plop, fizz fizz" for Alka-Seltzer and "I ♥ NY."

[42] Joe Ligo, "The Levi's AMC Gremlin Wasn't Just Quirky–It Fashioned a Movement," July 9, 2022, https://www.hagerty.com/media/car-profiles/the-levis-amc-gremlin-wasnt-just-quirky-it-fashioned-a-movement/.

but also a crisis of confidence in its identity as the world's automotive leader. Creating renewed fashion-driven hype for big American cars seemed like a reasonable play, because even in the 1970s, consumer behavior was not uniformly frugal, rational, or coherent. The decade witnessed the rise of lifestyle branding, as designer labels became shorthand for sophistication and cultural capital. Americans redefined consumption of branded consumer goods as a newly reconstituted form of self-expression. Automakers recognized an opportunity; if cars could no longer dominate purely on technology and engineering, perhaps they could remain desirable by connecting to new emerging expressions of personal style.

Lincoln's Designer Series embodied this logic in nearly every aspect. Introduced in 1976 for the Continental Mark IV, each version wore the branding of a celebrated fashion house and was carefully styled to appeal to consumer aesthetic desires and represent the design sensibilities of the respective fashion brands.[43] The Bill Blass edition incorporated deep navy paint with cream interiors to evoke nautical leisure and East Coast elegance. The Cartier edition, finished in soft Dove Gray, aimed to convey timeless refinement. Givenchy's Aqua Blue edition offered an understated chic, and Pucci's dramatic red-and-silver scheme exuded a theatrical flair. At least, that is what the manufacturers, the fashion brands, and the automotive designers hoped the public would see. It was, of course, a bait-and-switch: baiting consumers with color and style in exchange for less performance. The manufacturers hoped consumers would be willing to trade things desired in the past, like power, speed, and size, for branded status, the illusion of opulence, and exclusivity. As one would expect, the manufacturers made consumers pay up for the privilege. These accessory packages commanded significant price premiums, often $1500 to $2000 above the base model, which was around $11,000 in 1976. For the most part, these collaborative endeavors worked. Within months of their debut,

[43] MCG, "Rolling Fashion: The Continental Mark IV Designer Series," Mac's Motor City Garage, August 29, 2020, https://macsmotorcitygarage.com/rolling-fashion-the-continental-mark-iv-designer-series/; Bill Rothermel, "Designer Edition Continentals: Lincoln's Legacy Of Plush Luxury, Courtesy Of Celebrated Names In Style," *Hemmings.Com*, August 26, 2023, https://www.hemmings.com/stories/designer-edition-continentals/; Brett Berk, "Landau Yachts: The History of Lincoln's Designer Series," Features, *Autoblog*, February 6, 2022, https://www.autoblog.com/features/lincoln-designer-series-history; Corey Lewis, "Rare Rides Icons: The Lincoln Mark Series Cars, Feeling Continental (Part XX)," *Thetruthaboutcars.Com*, November 17, 2022, https://www.thetruthaboutcars.com/cars/rare-rides/rare-rides-icons-the-lincoln-mark-series-cars-feeling-continental-part-xx-44497576.

Designer Series models accounted for more than a quarter of Lincoln Mark IV sales.[44]

Cadillac would take the idea even farther. In 1978, it introduced the Cadillac Seville by Gucci, a limited-edition model festooned with Gucci's green-and-red striping, monogrammed upholstery, bespoke Italian leather interior, and an abundance of 24K gold plating inside and out, and to top it off, the car came with a matching Gucci luggage set.[45] Unlike Lincoln's relatively restrained approach, the Seville by Gucci was a full-bodied articulation of an international luxury brand blending its design acumen with an automotive brand known to produce the most luxurious automobiles. It appealed to consumers eager to signal not only wealth and cosmopolitan sophistication, but also those willing to display their Gucci affinity with anyone who saw them drive by. Produced in smaller numbers than Lincolns, the Seville by Gucci underscored how deeply intertwined fashion and automobiles had become by the end of the decade.

The allure of couture cars may lie less in their technological and mechanical distinction than in their performative and symbolic value. In moments of economic crisis, like 1973, the desire to perform expressions of distinguishing taste, affluence, and cosmopolitanism becomes more desirable in capitalist economies. It did not hurt that Lincoln, Cadillac, and nearly all the other manufacturers made sure they produced the models in limited quantities to reinforce their exclusivity and desirability.[46] In an era of economic instability, American manufacturers bet that exclusivity, for certain models, would be more valuable than efficiency. Even as the business of manufacturing automobiles moved to smaller fuel-efficient cars, some buyers and executives still valued the older aesthetics. For instance, the Lincoln Mark IV, manufactured from 1971–1976 and the Lincoln Mark V, produced from 1977–1979, were some of the longest coupes Lincoln ever produced, and among the longest two-door cars ever

[44] Berk, "Landau Yachts."

[45] Jonathan Wells, "Is the Cadillac Seville 'Gucci Edition' the Most Stylish Car in History?," *Gentleman's Journal*, n.d., accessed January 18, 2026, https://www.thegentlemansjournal.com/article/cadillac-seville-gucci-edition-car-about/.

[46] Corey Lewis, "Rare Rides Icons: The Lincoln Mark Series Cars, Fee ling Continental (Part XXVIII)," *Thetruthaboutcars.Com*, January 20, 2023, https://www.thetruthaboutcars.com/cars/rare-rides/rare-rides-icons-the-lincoln-mark-series-cars-feeling-continental-part-xxviii-44498890.

made.[47] This specialness helped quell some of the criticism of broader quality issues for the brand and the American automotive industry as well. At a time when American cars were derided for poor fuel efficiency and declining workmanship, couture editions offered up luxury and refinement as useful and, for some, an effective counter-narrative.

Couture collaborations also influenced automotive aesthetics. Automotive designers borrowed cues from fashion principles to modify and develop textiles and colors to adorn automotive interiors and exteriors. The AMC Pucci editions mirrored the vibrant colors seen in prints in their runway collections. Cartier models deployed subtle earth tones to echo the jewelry house's timeless simplicity and minimalist design language. The ultimate goal was to enhance, refine, augment, and extend a driver's visual landscape within the car. Lincoln specifically intended for the interior experience to be the polar opposite of the experience in a Honda Civic or Toyota Corolla. The idea was to redefine and differentiate the driving experience by providing a set of new luxury-derived sensations. Loud and engaged driving is a hallmark of a performance car.

Silent opulence is a marker of a luxury vehicle. The couture cars wanted drivers to experience being dressed by one of the best clothing designers in the world and to prove that your personal style did not end with the clothes draping your torso but extended into the vehicle you chose to drive. The collaborations legitimized cars as a renewed type of cultural artifact. If couture fashion represented the apex of design and luxury, then couture cars positioned American automobiles as objects of artistry, well beyond functional transportation and practical engineering. This rebranding was crucial in sustaining desirability at a time when Japanese competition eroded Detroit's functional and practical advantages.

High Values of One Plus High Values of One Equals Three

While teaching at the University of Illinois over a decade ago, I had a wonderful colleague in the College of Engineering, Michael Loui. I would occasionally guest lecture in his course on engineering ethics and in one of those sessions, I heard him say, "high values of one plus high values of one

[47] Matt Nelson, "The Longest Coupe Lincoln Ever Built Can Be Yours For Used Civic Money," CarBuzz, November 29, 2025, https://carbuzz.com/the-longest-lincoln-coupe-ever-built-can-be-yours-for-used-civic-money/.

equals three" to make a point about perspective. I assume this may be a common phrase heard by engineering undergraduates across the world, but it was new to me. Honestly, I didn't get it at first, so I had to ask for clarification. He would explain that 1.4, which is a "high" value of 1, rounds down to 1. But 1.4 plus 1.4 equals 2.8, which rounds up to 3. This was revelatory. I assume that was the point Michael was trying to make to a room full of undergraduate engineers, but I was probably the only one in the classroom whose mind spun off in multiple different directions. For that little kernel, I thank Michael.

That little nugget reminds me of the brilliance of Ron's work to take a seemingly simple question or observation and have it open a new door and reveal an entirely new world of understanding. I think that is the mark of transformative scholarship. It is easy to dazzle with the linguistic gymnastics familiar in academic work, but it is even more difficult to enlighten through careful, thoughtful, and rigorous work that elicits new meanings from the simplest, deeply observant observations. This is what this essay attempted to model.

It also tells a technological innovation story that did not begin with patent specifications or intellectual property rights. It's a culturally situated design innovation story where the innovation was aesthetic. It's an innovation story, not of technical innovation, but a design innovation of a technology, an innovation that predates Apple's introduction of the seductive, colorful, translucent enclosures of the iMac G3 in 1998.

Taken together, these collaborations foreshadow today's culture of cross-branding, from celebrity sneaker lines to designer everything. Couture cars were an early experiment in merging industries to sell identity as much as product. Currently embraced as a curious artifact of 1970s automotive culture, not all observers were convinced at the time. Some journalists dismissed couture cars as superficial exercises in branding and upselling logos and paint as limited-edition packages. Presently, well-preserved Designer Series Lincolns and Sevilles by Gucci are prized artifacts of 1970s consumer culture and style, valued not for their horsepower but for their cultural symbolism. The persistent collector interest suggests these critiques underestimated the cultural resonance of couture cars.

Couture cars emerged in the 1970s as a bold experiment to reimagine the automobile as a luxury accessory as a response to an age of political, cultural, and economic uncertainty partially created by the global business of oil. By partnering with fashion brands such as Bill Blass, Cartier, Givenchy, Pucci, and Gucci, American automakers reframed automobiles as lifestyle artifacts, aligning them with the prestige of haute couture. These collaborations enhanced desirability through exclusivity, aesthetic innovation, and cultural association, even as Detroit struggled with broader structural challenges that persist. In retrospect, couture cars were more than marketing novelties. They were part of a broader cultural shift in which consumption became a vehicle for self-expression and identity. In the fusion of fashion and automobiles, American automobile manufacturers contributed to building a cultural and aesthetic world in which branding would be as important as engineering, in a world that, for better or worse, defines consumer culture today.

Bibliography

Alhajji, A. F. "Three Decades After the Oil Embargo: Was 1973 Unique?" *The Journal of Energy and Development* 30, no. 2 (2005): 223–37.

Barber, Brad M., and Masako N. Darrough. "Product Reliability and Firm Value: The Experience of American and Japanese Automakers, 1973–1992." *Journal of Political Economy* 104, no. 5 (1996): 1084–99.

Bennett, Ricard. "The 1979 Lincoln Continental Mark V Was a Designer's Delight." May 22, 2019. https://www.hagerty.com/media/opinion/1979-lincoln-continental-mark-v-designers-delight/

Berk, Brett. "Landau Yachts: The History of Lincoln's Designer Series." Features. *Autoblog*, February 6, 2022. https://www.autoblog.com/features/lincoln-designer-series-history

Eden, Lorraine, and Maureen Appel Molot. Made in America? The US Auto Industry, 1955–95 *The International Executive* 38, no. 4 (1996): 501–41. https://doi.org/10.1002/tie.5060380406.

Flink, James J. "Three Stages of American Automobile Consciousness." *American Quarterly* 24, no. 4 (1972): 451–73. https://doi.org/10.2307/2711684.

Fouché, Rayvon. *Black Inventors in the Age of Segregation.* Johns Hopkins University Press, 2005. https://doi.org/10.56021/9780801873195.

Fouché, Rayvon. *Game Changer: The Technoscientific Revolution in Sports.* Johns Hopkins University Press, 2017.

Fouché, Rayvon. "Say It Loud, I'm Black and I'm Proud: African Americans, American Artifactual Culture, and Black Vernacular Technological Creativity." *American Quarterly* 58, no. 3 (2006a): 639–61.

Fouché, Rayvon. "The Wretched of the Gulf: Racism, Technological Dramas, and Black Politics of Technology." *The Black Scholar* 36, no. 4 (2006b): 7–12.

Friedlaender, Ann F., Clifford Winston, and Kung Wang. "Costs, Technology, and Productivity in the U.S. Automobile Industry." *The Bell Journal of Economics* 14, no. 1 (1983): 1–20. https://doi.org/10.2307/3003534.

Godshall, Jeffrey I., James K. Wagner, and James A. Wren. The Automobile - Unwanted Technology: Part II the Middle Years (1930–1960). *SAE Transactions* 100, no. 6 (1991): 1353–58. https://doi.org/10.4271/910903.

Graf, Rüdiger. "Making Use of the 'Oil Weapon': Western Industrialized Countries and Arab Petropolitics in 1973–1974." *Diplomatic History* 36, no. 1 (2012): 185–208.

Hall, Laura. "The Evolution of CAFE Standards: Fuel Economy Regulation Enters Its Second Act." *Transportation Law Journal* 39, no. 1 (2011): 1–30.

Hervey, Jack L. 1994. *The 1973 Oil Crisis: One Generation and Counting*. Chicago Fed Letter 86. https://www.chicagofed.org/publications/chicago-fed-letter/1994/october-86

Jacobs, Meg. *Panic at the Pump: The Energy Crisis and the Transformation of American Politics in the 1970s*. Hill and Wang, 2016.

Jensen, Tom. "Win on Sunday, Sell on Monday." NASCAR Hall of Fame, June 23, 2023. https://www.nascarhall.com/blog/win-on-sunday-sell-on-monday

James A. Dunn. *Driving Forces: The Automobile, Its Enemies, and the Politics of Mobility*. Brookings Institution Press, 1998.

Kandiyoti, Rafael. *Pipelines: Flowing Oil and Crude Politics*. I.B.Tauris, 2008.

Kaster, James. 2011. *American Automotive Design Trends / The Couture Car: High Style for High Society*. Lulu.com.

Kawahara, Akira. *The Origin of Competitive Strength: Fifty Years of the Auto Industry in Japan and the U.S.* Reprint of the original. Springer. 2012.

Kline, Ronald, and Trevor Pinch. "Users as Agents of Technological Change: The Social Construction of the Automobile in the Rural United States." *Technology and Culture* 37, no. 4 (1996): 763–95.

Kline, Ronald R. *Consumers in the Country: Technology and Social Change in Rural America*. Johns Hopkins University Press, 2000.

Lewis, Corey. "Rare Rides Icons: The Lincoln Mark Series Cars, Feeling Continental (Part XX)." *Thetruthaboutcars.Com*, November 17, 2022. https://www.thetruthaboutcars.com/cars/rare-rides/rare-rides-icons-the-lincoln-mark-series-cars-feeling-continental-part-xx-44497576

Lewis, Corey. "Rare Rides Icons: The Lincoln Mark Series Cars, Feeling Continental (Part XXVIII)." *Thetruthaboutcars.Com*, January 20, 2023. https://www.thetruthaboutcars.com/cars/rare-rides/rare-rides-icons-the-lincoln-mark-series-cars-feeling-continental-part-xxviii-44498890

Ligo, Joe. "The Levi's AMC Gremlin Wasn't Just Quirky–It Fashioned a Movement." July 9, 2022. https://www.hagerty.com/media/car-profiles/the-levis-amc-gremlin-wasnt-just-quirky-it-fashioned-a-movement/

Liu, Chen, William Rouse, and Zhongyuan Yu. "When Transformation Fails: Twelve Case Studies in the American Automobile Industry." *Journal of Enterprise Transformation* 5, no. 2 (2015): 71–112. https://doi.org/10.1080/19488289.2015.1019654.

MCG. "Rolling Fashion: The Continental Mark IV Designer Series." Mac's Motor City Garage, August 29, 2020. https://macsmotorcitygarage.com/rolling-fashion-the-continental-mark-iv-designer-series/

Nelson, Matt. "The Longest Coupe Lincoln Ever Built Can Be Yours For Used Civic Money." CarBuzz, November 29, 2025. https://carbuzz.com/the-longest-lincoln-coupe-ever-built-can-be-yours-for-used-civic-money/

Pierce, John R. "The Fuel Consumption of Automobiles." *Scientific American* 232, no. 1 (1975): 34–47.

Rome, Adam. *The Genius of Earth Day: How a 1970 Teach-In Unexpectedly Made the First Green Generation*. Hill and Wang, 2013.

Rose, Mark H. *Interstate: Express Highway Politics 1939–1989*. University of Tennessee Press, 1990.

Rothermel, Bill. "Designer Edition Continentals: Lincoln's Legacy Of Plush Luxury, Courtesy Of Celebrated Names In Style." *Hemmings.Com*, August 26, 2023. https://www.hemmings.com/stories/designer-edition-continentals/

Sachs, Wolfgang. 1992. *For Love of the Automobile Looking Back into the History of Our Desire*. Trans. Don Reneau. University of California Press.

Schleicher, David. From Here All-the-Way-Down, or How to Write a Festschrift Piece." *Tulsa Law Review* 48, 3 (2013): 401–25.

Sheller, Mimi. "Automotive Emotions: Feeling the Car." *Theory, Culture & Society* 21, 4–5 (2004): 221–42. https://doi.org/10.1177/0263276404046068.

Sheth, Jagdish N., and Atul Parvatiyar. "The Evolution of Relationship Marketing." *International Business Review* 4, 4 (1995): 397–418. https://doi.org/10.1016/0969-5931(95)00018-6.

Sonfield, Matthew C. "Custom Automotive Coachbuilding in the United States, 1900–1940." *Design Issues* 12, 2 (1996): 47–60. https://doi.org/10.2307/1511713.

Stein, Jonathan A., ed. *The Art and Colour of General Motors*. Coachbuilt Press, 2008.

TIME. "U.S. Business: Irreverence at American." September 22, 1967. https://web.archive.org/web/20110123111601/http:/www.time.com/time/magazine/article/0,9171,837337,00.html

Tomeh, George. "Oapec: Its Growing Role in Arab and World Affairs." *The Journal of Energy and Development* 3, no. 1 (1977): 26–36.

Vázquez-Fariñas, María. "Major Economic Recessions in the Last Quarter of the 20th Century: The Oil Crisis (1973–1980)." In *The Age of Global Economic Crises*, edited by Juan Manuel Matés-Barco and María Vázquez-Fariñas. Routledge, 2023.

Venn, Fiona. *The Oil Crisis*. Routledge, 2002.

Wells, Jonathan. "Is the Cadillac Seville 'Gucci Edition' the Most Stylish Car in History?" *Gentleman's Journal*, n.d. Accessed January 18, 2026. https://www.thegentlemansjournal.com/article/cadillac-seville-gucci-edition-car-about/

CHAPTER 4

Accommodating Urban Micromobility Innovations

Clinton J. Andrews and Lina Moe

Introduction

Every city brings together private and public resources, and this combination makes the city viable. Private actors own most buildings, enterprises, and technological systems, while public institutions typically take responsibility for providing infrastructure and certain services, and setting rules. Urban governance involves both types of actors, and more, who coordinate using approaches spanning integrated regimes, looser relational networks, aspirational associative democracy, and plain old pluralism.[1]

[1] Susan S. Fainstein et al., eds., *Restructuring the City: The Political Economy of Urban Redevelopment* (Longman, 1983). Karen Mossberger and Gerry Stoker, "The Evolution of Urban Regime Theory: The Challenge of Conceptualization," *Urban Affairs Review* 36, no. 6 (2001): 810–35, https://doi.org/10.1177/10780870122185109. Luuk Boelens, "Theorizing Practice and Practising Theory: Outlines for an Actor-Relational-Approach in

C. J. Andrews • L. Moe (✉)
Rutgers University, New Brunswick, NJ, USA
e-mail: lina.moe@rutgers.edu

J. L. Lieberman, A. Mohsin (eds.), *Technological Humanities*,
https://doi.org/10.1007/978-3-032-24741-4_4

Urban technological innovation introduces coordination challenges at the private-public interface. Such innovations often originate in the private sector, yet their deployment depends on infrastructure and rules provided by the public sector. This public-private interdependence poses a policy question: is it better to build infrastructure ahead of demand or in response to it? Albert Hirschman famously argued that it is thriftier to build infrastructure in response to demand growth, but engineers and planners worldwide prefer to build ahead of demand in order to ensure an acceptable quality of service for users.[2] This tension also applies to cases in which new technologies require modifications to existing infrastructures and rules.

This chapter considers how cities are attempting to resolve the react-vs-anticipate tension in the case of emerging micromobility technologies such as electric scooters and electric bikes. Our framing of the problem diverges from recent literature that views urban micromobility as primarily a regulation problem featuring contestation between market actors and regulatory actors, or a technical problem of whether it is more sustainable than other mobility technologies.[3] Instead, we approach it as an innovation policy problem: how can cities create governance and infrastructural conditions that accommodate technological disruption and shape it toward public goals? Cities may tolerate the technological "push" of platform-enabled mobility services if it appears alongside a "pull" of municipal objectives such as reducing greenhouse gas emissions, filling first- and last-mile gaps, and improving transport equity. We argue that the distinctive contribution of an innovation policy lens is its emphasis on adaptive capacity or the ways in which cities build regulatory, informational, and organizational systems capacity to absorb technological shocks and redirect them.

Planning," *Planning Theory* 9, no. 1 (2010): 28–62, https://doi.org/10.1177/1473095209346499. Joseph P. Viteritti, "The City and the Constitution: A Historical Analysis of Institutional Evolution and Adaptation," *Journal of Urban Affairs* 12, no. 3 (1990): 221–36, https://doi.org/10.1111/j.1467-9906.1990.tb00215.x.

[2] Albert O. Hirschman, *Development Projects Observed* (Brookings Institution Press, 2011).

[3] Clare Field and Ihnji Jon, "E-Scooters: A New Smart Mobility Option? The Case of Brisbane, Australia," *Planning Theory & Practice* 22, no. 3 (2021): 368–96, https://doi.org/10.1080/14649357.2021.1919746. Hélie Moreau et al., "Dockless E-Scooter: A Green Solution for Mobility? Comparative Case Study between Dockless E-Scooters, Displaced Transport, and Personal E-Scooters," *Sustainability* 12, no. 5 (2020): 1803, https://doi.org/10.3390/su12051803.

Building on Hirschmann's demand-driven infrastructure strategy and associated "hiding hand" concept, and Ron Kline's user-driven innovation, we propose that recent technological disruptions in the form of micromobility options on urban streets pushed cities to consider rapid, lightweight forms of adaptive governance, which we term "soft infrastructure." We situate these policy responses to technological disruption in the broader contexts of the political economy of platform and gig economies, as well as urban mobility plans that prioritize expanding transportation access while reducing congestion and carbon emissions. In this way, the e-scooter episode becomes not only a transportation story but a window into how local governments navigate the tradeoffs and uncertainties inherent in regulating emergent technologies. Our case study investigation of Santa Monica—the first site of the surprise arrival of e-scooters—demonstrates that adaptive governance involves not only setting guardrails for technology companies that seek regulatory loopholes but also injecting city governance with toleration for a degree of disruption in the service of new solutions to the challenge of urban mobility.

Urban Technological Innovation

New technology diffusion depends on both intrinsic and relational characteristics of the innovation, including relative advantage, compatibility with existing systems, complexity, observability, and trialability.[4] But relational characteristics such as relative advantage and compatibility depend in part on the infrastructure and rules of the host jurisdiction.

The capability and willingness of the public actors to facilitate adoption of the private innovation can be determinative. Some jurisdictions are more flexible than others, and, importantly, some encourage the possibility of learning by trial and error, whether through disjointed incrementalism or explicit experimentation.[5] Transition management advocates, in particular, call for more systematic experimentation in the pursuit of a sustainable future.[6] They favor conducting limited experiments within

[4] Everett M. Rogers, *Diffusion of Innovations*, Fifth edition (Free Press, 2003).

[5] Charles E. Lindblom, "The Science of 'Muddling Through,'" *Public Administration Review* 19, no. 2 (1959): 79, https://doi.org/10.2307/973677. Sebastian Heilmann, "Policy Experimentation in China's Economic Rise," *Studies in Comparative International Development* 43, no. 1 (2008): 1–26, https://doi.org/10.1007/s12116-007-9014-4.

[6] Frank W. Geels and Johan Schot, "Typology of Sociotechnical Transition Pathways," *Research Policy* 36, no. 3 (2007): 399–417, https://doi.org/10.1016/j.respol.2007.01.003.

protected niches and incorporating what is learned into broader deployments. Urban technologies are especially amenable to experiments because only one city or neighborhood at a time needs to be put at risk, and lessons learned in one location can inform subsequent technology developments and public policies. Cities, as much as states, serve as democracy's laboratories.[7]

Intrinsic technology characteristics, relational factors, jurisdictional openness to experimentation, user decisions, and existing social practices each play roles in determining which technological changes come to cities. Many of these factors have important economic, political, social, and psychological dimensions. Hirschman's simple question opens a Pandora's Box of complexity.

Hirschman introduced the "hiding hand" principle to explain a paradox of planning and creativity.[8] In his fieldwork he found that when planners embarked on big infrastructure projects, they often underestimated the true complexity, cost, and difficulties that lay in store. These technological, economic, or social hurdles were "hidden" at the project's outset. However, once underway, when previously unseen problems inevitably arose, people would be more likely to tenaciously improvise solutions and innovate new ways forward rather than give up. This hard work would often generate better outcomes than if planners had sought to anticipate every issue from the start. Unanticipated problems, Hirschman posited, catalyzed real-time problem-solving, but as importantly, ignorance offered a license for projects to get out of the gate. By contrast, full awareness of obstacles would discourage innovators from ever taking on ultimately beneficial projects.

New technology adoption depends on more than just the technology provider and the host jurisdiction because the users decide whether a new technology offers relative advantage in their lives. User-friendliness, or usability, depends on technology characteristics, user characteristics, contextual factors (partly determined by the host jurisdiction), and the task at

[7] F.A.O Schwartz, "Opening Remarks: The States and Cities as Federal Laboratories of Democracy," *Fordham Urb. L.J.* 27 (1999), https://ir.lawnet.fordham.edu/ulj/vol27/iss1/7.

[8] Hirschman 1967/2011.

hand.[9] The innovation process itself relies on users for feedback that iteratively shapes what gets deployed.[10]

In *Consumers in the Country* (2000), Kline picks up this creative tension in the history of telecommunications networks when the disregard of rural residents and ignorance of the obstacles to connecting the vast countryside of the U.S. catalyzed local innovation, as users creatively accommodated new technology and ingeniously adapted it to their needs.[11] When telephone companies underestimated the challenge and cost of serving rural areas, essentially leaving farmers to fend for themselves, these rural actors first had to decide if they wanted to be part of a communication network and then, once committed, they had to find creative solutions, such as barbed wire fences and home-built switchboards, to connect isolated homesteads to a network that had not been designed with them in mind. Similarly, farmers who bought early automobiles discovered these machines were not adapted to an agricultural setting (pulling heavy loads down muddy roads) but could be transformed into a multi-purpose farm tool by removing the rear wheels of a Model T and attaching belt drives to power threshing machines or to turn a washing machine drum.

Users live in specific social contexts and new technologies need to fit into or productively disrupt extant social practices to achieve widespread use. Social practices depend equally on the material dimension (technology), the competencies of the users, and the social meanings attached to the practice.[12] Unbundling these elements and re-bundling with new technologies yield new social practices, often a slow process. Not only must the user learn new skills, but considerations of desirability, status, reputation, and even legality may slow technology adoption.

Kline and Pinch's analysis of the automobile remains a foundational study in how users are also innovators, or how when technology pushes,

[9] Nigel Bevan and Miles Macleod, "Usability Measurement in Context," *Behaviour & Information Technology* 13, nos. 1–2 (1994): 132–45, https://doi.org/10.1080/01449299408914592.

[10] Eric Von Hippel, *Democratizing Innovation* (The MIT Press, 2005), https://doi.org/10.7551/mitpress/2333.001.0001.

[11] Ronald R. Kline, *Consumers in the Country: Technology and Social Change in Rural America*, Revisiting Rural America (Johns Hopkins university press, 2000).

[12] Elizabeth Shove et al., *The Dynamics of Social Practice: Everyday Life and How It Changes* (SAGE Publications Ltd, 2012), https://doi.org/10.4135/9781446250655.

society pushes back.[13] America became a "car culture" not simply because of the introduction of the automobile but because of how people used cars, sometimes creating new technologies from these machines and more frequently inventing new forms of life that took advantage of these inventions. Who's to say what a micromobility culture could look like?

In this way Kline was a keen observer of the dialectic between technical innovation and social context, showing how ordinary people, as well as the local character of institutions, influence the course of new technologies. Kline was quick to point out that narratives of technological innovation are not exclusively urban, a bias that persists today in the disproportionate focus on innovation in "superstar cities." Such metropolitan bias continues to overshadow meaningful innovations in smaller urban areas and, even more so, in sparsely populated rural regions like the northern Wisconsin backwoods or Iowa hill country that Kline closely examined. As Kline's *Steinmetz* (1992) showed, socio-technical governance is a product of local context, as evidenced by the entwined fates of General Electric and early twentieth-century Schenectady.[14]

Hirschman often studied large infrastructure projects, such as dams, highways, and irrigation systems, and Kline examined major inventions of the twentieth century—the car and the telecommunications system—which often demanded centralized bureaucratic responses to mobilize large-scale land transformation. Micromobility innovation offers a new, more compact lens for contemplating the evolution of Hirschman's "hiding hand" and Kline's ideas about local socio-technical environments. Micromobility companies scattered a decentralized, rapidly deployable technology across the urban landscape—the technological push—without involvement, investment, or permission from local governments or planners.

The technological challenge that launched a thousand angry public comments was not defined by the sheer magnitude or complexity of a physical engineering project but rather the complexity of regulating small-scale, dispersed operations and managing unforeseen social consequences at a dauntingly granular level. Cities were forced to decide if the

[13] Ronald Kline and Trevor Pinch, "Users as Agents of Technological Change: The Social Construction of the Automobile in the Rural United States," *Technology and Culture* 37, no. 4 (1996): 763–95.

[14] Ronald R. Kline, *Steinmetz: Engineer and Socialist*, Johns Hopkins Studies in the History of Technology, new ser., no. 13 (Johns Hopkins University Press, 1992).

technological irritant was worth responding to, and if so, what kind of governance and infrastructure building would be necessary to create a soft landing for this insurgent technology.

Micromobility

Mobility is an important and problematic urban function that exemplifies the need for public-private coordination. Pedestrians, draft animals, bicycles, cars, and buses depend on public streets and sidewalks; and rail-based systems, whether above or below ground, require dedicated rights-of-way. Multi-mode travel is nothing new in lower-income countries where automobiles have not displaced other modes. Chaotic scenes of stubborn donkeys, ringing bike bells, whining motorbikes, swerving auto-rickshaws, honking taxi horns, and lumbering buses and trucks all sharing the common roadway exemplify transportation norms in many parts of the world. However, where cars have taken over, this diversity has shrunk, increasing aggregate environmental impacts and reducing safety for those not in cars. For more than a century, many cities have struggled to serve pedestrians, cars, mass transit, and other vehicles well, with cars often taking priority for funding and street space.

Recent years have brought new micromobility options, defined here as lightweight vehicles including bicycles, scooters, skateboards, and electric-motor-powered versions of each, that typically operate at speeds below 25 km/h (15 mph), may be shared or privately owned, and are designed to carry humans.[15] Additional mobility technologies fall just outside this definition, such as seated electric scooters, electric wheelchairs, and motor bikes powered by internal combustion engines. Many micromobility technologies could potentially enhance the personal mobility of people who do not have ready access to cars or mass transit because they are low-income earners, disabled, elderly, or children. These technologies could also reduce the environmental impacts of transportation—the largest source of U.S. greenhouse gas emissions.[16] Early studies of micromobility pilots show that some of these goals have been realized more than others;

[15] Dana Yanocha and Mackenzie Allan, *Electric Assist: Leveraging E-Bikes and E-Scooters for More Livable Cities* (Institute for Transportation & Development Policy, 2021).

[16] Environmental Protection Agency (EPA), U.S. 2021, *Fast Facts: U.S. Transportation Sector GHG Emissions*, https://www.epa.gov/greenvehicles/fast-facts-transportation-greenhouse-gas-emissions.

in particular, though scooter rides do appear to be replacing car trips, low-income users make up only a small percentage of scooter riders.[17]

Micromobility has a long technological history, spanning invention of the bicycle in 1817, e-bike in 1895, motorized scooter in 1916, and electronic Segway in 1997.[18] Modern e-bikes evolved independently in North America, Europe, and Asia during the 1990s, but they first achieved widespread usage in China in the early 2000s, reaching annual sales of 7.5 million units by 2004.[19] Modern e-scooters were first deployed at scale on a shared basis in the U.S. in 2017, due to "the intersection of multiple technological trends — the falling price of batteries and GPS trackers, the near-ubiquity of smartphones, and the rising demand for space in central cities," plus access to venture capital.[20]

Business models of shared micromobility have evolved through several generations including dockless free community bikeshare (Amsterdam in 1965); coin-operated community bikeshare (Copenhagen in 1995); docked bikeshare requiring a credit card (Lyon in 2005); docked e-bikeshare (Knoxville, TN, in 2011); dockless bikeshare with embedded GPS and requiring a credit card (Beijing in 2015); dockless e-bikes (Dallas, TX, in 2017); and dockless e-scooters (Santa Monica, CA, in 2017).[21]

[17] City of Santa Monica, *Shared Mobility Pilot Program Summary Report* (2019), https://www.smgov.net/uploadedFiles/Departments/PCD/Transportation/SantaMonicaSharedMobilityEvaluation_Final_110419.pdf.

[18] Frank Berto, "Bicycle," in *Britannica*, accessed November 16, 2025, https://www.britannica.com/technology/bicycle. Ogden Bolton, Electric Bicycle, U.S. Patent and Trademark Office Patent U.S. Patent No. 552,271, filed 1985. A.E.C. Gibson, Self-propelled Vehicle, U.S. Patent and Trademark Office Patent U.S. Patent No. 1,192,514, issued 1916. Dean L. Kamen et al., Human transporter, United States Patent US5701965A, filed May 27, 1994, and issued December 30, 1997, https://patents.google.com/patent/US5701965A/en. Sarah Holder, "The Man Behind the Scooter Revolution," *Bloomberg. Com*, September 26, 2018, https://www.bloomberg.com/news/articles/2018-09-26/how-a-kids-scooter-became-a-micro-mobility-revolution.

[19] Peter Fairley, "China's Cyclists Embrace Electric Bikes - IEEE Spectrum," *IEEE Spectrum*, June 1, 2005, https://spectrum.ieee.org/chinas-cyclists-take-charge.

[20] Matthew Yglesias, "The Electric Scooter Boom — and Cities' Misguided Attempts to Squash It — Explained," *Vox*, September 10, 2018, https://www.vox.com/the-goods/2018/9/10/17631318/electric-scooters-bird-city-regulations-sustainability.

[21] Paul DeMaio, "Bike-Sharing: History, Impacts, Models of Provision, and Future," *Journal of Public Transportation* 12, no. 4 (2009): 41–56, https://doi.org/10.5038/2375-0901.12.4.3. Susan A. Shaheen et al., "Bikesharing in Europe, the Americas, and Asia: Past, Present, and Future," *Transportation Research Record: Journal of the Transportation Research Board* 2143, no. 1 (2010): 159–67, https://doi.

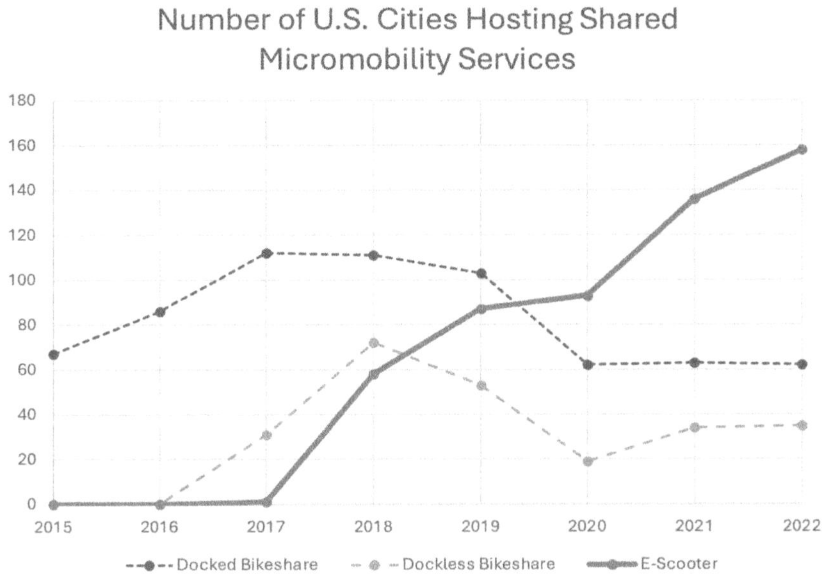

Fig. 4.1 Number of U.S. cities hosting shared micromobility services from 2015 to 2022 (Data Source: BTS 2024)

Figure 4.1 shows the number of U.S. cities hosting shared micromobility services from 2015 to 2022, confirming the pattern of docked bikeshare deployments, followed by dockless bikeshare, and then shared e-scooters. This ordering is also evident in individual cities.

Current e-scooter deployments thus build on a historical base of related technological and business innovations. Global e-scooter sales in 2020 are estimated at USD 19.4 billion, of which 74% are in the Asia-Pacific region

org/10.3141/2143-20. Brian Casey Langford et al., "North America's First E-Bikeshare: A Year of Experience," *Transportation Research Record: Journal of the Transportation Research Board* 2387, no. 1 (2013): 120–28, https://doi.org/10.3141/2387-14. Christopher Moon et al., *The Evolution of Bike Sharing*, January 14, 2019, https://www.wri.org/research/evolution-bike-sharing. Bureau of Transportation Statistics, "Bikeshare (Docked and Dockless) and E-Scooter Systems by Year and City Served," U.S. Department of Transportation, November 29, 2024, https://data.bts.gov/Bicycles-and-Pedestrians/Bikeshare-Docked-and-Dockless-and-E-scooter-System/cqdc-cm7d/about_data.

and 7% are in the U.S.[22] Grandview's report states that sealed lead-acid batteries still dominate (60% market share in 2020), but lighter lithium-ion batteries are the most rapidly growing battery type. Within the U.S., there has been exponential growth in shared e-scooter trips, from close to 0 in 2017, to 38.5 million in 2018, and 88.5 million in 2019, before crashing due to the COVID-19 pandemic, before which e-scooter trips exceeded bikeshare trips by nearly a factor of 2x.[23] Ridership data from 2021 show a significant rebound, above 2019 levels.[24]

Micromobility's expansion has been complexly intertwined with the rise of the gig economy workforce, with for-hire independent contractors used to provide core services for companies in the growing so-called sharing economy. Early operators relied heavily on gig labor for scooter maintenance, recharging, and fleet redistribution. This reliance on informal, precarious employment created tensions with municipal regulators, leading some cities, such as New York, to explicitly require operators to move away from gig labor models toward more formal employment practices.[25]

Additionally, scooter fleets arrived as cities were incorporating emissions reductions and climate pledges into master plans. Some cities have tolerated challenges for the promise of a new mobility form with reduced emissions. Yet, the promise of an equitable low-carbon alternative remains partly unmet, with scooters attracting disproportionately younger, higher-income, male riders.

The e-scooter's global diffusion has not been smooth. They are ubiquitous in some cities and shut out of others. In some places where pilot deployments were initially welcomed, they have not been invited back. Some e-scooter-share companies have gone bankrupt or merged. Perceptions that e-scooters are unsafe and add street clutter have become widespread. Why have they succeeded in some places but not others?

[22] Grandview Research, *Electric Scooters Market Size & Share Analysis Report, 2030*, GVR-1-68038-196-2 (2021), https://www.grandviewresearch.com/industry-analysis/electric-scooters-market.

[23] National Association of City Transportation Officials, *Shared Micromobility in the U.S* (2020), https://nacto.org/wp-content/uploads/2020/08/2020bikesharesnapshot.pdf.

[24] *Bikeshare Ridership Exceeds Pre-Pandemic* (Bureau of Transportation Statistics, 2021), https://www.bts.gov/newsroom/bikeshare-ridership-exceeds-pre-pandemic-2019-high-fourth-straight-month.

[25] John Seabrook, "The E-Scooters Loved by Silicon Valley Roll into New York," *New Yorker*, April 19, 2021, https://www.newyorker.com/magazine/2021/04/26/the-e-scooters-loved-by-silicon-valley-roll-into-new-york.

Case Studies

Case study research is a useful method for examining the deployment of e-scooters in cities because it highlights contextual aspects of the technology diffusion story. The congestion on both roads and sidewalks associated with urban cores makes e-scooter deployments more difficult, but also more attractive than in many other contexts. We know in advance that there is variation in outcomes, so it is reasonable to choose contrasting cases and seek to understand how the differences emerged. Major cities illustrate these differences:

- Paris in 2018 welcomed shared e-scooters but in 2023 prohibited them after several fatalities. Private e-scooters are still allowed.[26]
- New York City in 2020 (Local Law 74) prohibited shared e-scooters in Manhattan but allowed a pilot in the Bronx, which in 2024 expanded to Queens. Private e-scooters are also allowed.[27]
- London initially prohibited shared e-scooters in most areas of the city but allowed pilots in selected neighborhoods starting in 2021; now their programs are expanding. Private e-scooters are still prohibited.[28]
- Tokyo in 2022 reclassified e-scooters to allow their use without a special driver's license, and both shared and private e-scooters are now ubiquitous.[29]

The small city of Asbury Park, New Jersey, chose a government-led route to shared micromobility. This is a small tourism destination that sits close to major East Coast population centers. Following a successful experience with bikeshare programs, in 2019 the city sought out shared e-scooter vendors to serve some of the mobility needs of the thousands of daily visitors to the city's boardwalk and beaches. The city managed each

[26] *Paris' Rental Electric Scooter Ban Has Taken Effect*, directed by Eleanor Beardsley, NPR, September 1, 2023, https://www.npr.org/2023/09/01/1197329094/paris-rental-electric-scooter-ban-has-taken-effect. European Consumer Centre Germany (ECCG), *Bike and E-Scooter Hire in Paris*, April 2024, https://www.evz.de/en/travelling-motor-vehicles/e-mobility/two-wheelers/bike-and-scooter-rental-paris.htm.

[27] NY DOT, *Shared E-Scooter Program*, 2025, https://nycdotscootershare.info/faq.

[28] Transport for London, *Electric Scooters*, n.d., https://tfl.gov.uk/modes/driving/electric-scooter-rental-trial.

[29] Japan Dev Team (JVT), *Electric Scooters in Japan*, 2024, https://japan-dev.com/blog/electric-scooters-in-japan.

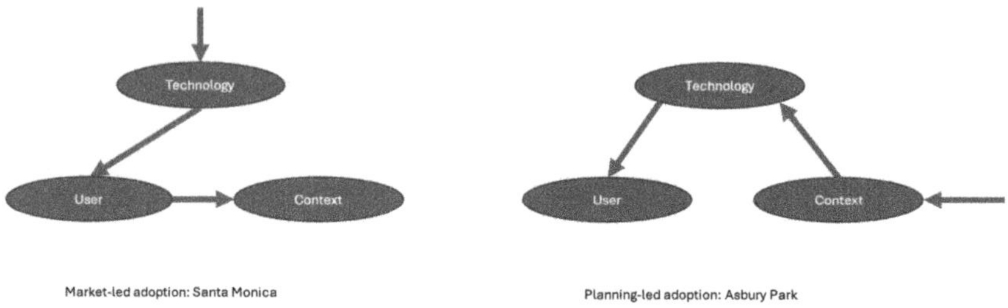

Fig. 4.2 Market-led adoption versus planning-led adoption

step in the deployments, from selecting vendors based on bids, to establishing clear deployment rules and policy priorities. Both shared and private e-scooter use continues today. However, the introduction of micromobility options has not always unfolded seamlessly. We turn to one case in depth to highlight important dimensions of the micromobility experiment, the first city where the e-scooters suddenly appeared one night in 2017: Santa Monica (Fig. 4.2).

Santa Monica Case Study

Santa Monica, a coastal city of roughly 90,000 in Los Angeles County, became a testing ground for micromobility when startup Bird suddenly dropped electric scooters on its streets in September 2017. At the time, no local laws or permits existed for these dockless, app-rented e-scooters. This unpermitted launch by Bird—soon joined by Lime—effectively turned the city of Santa Monica into a laboratory for a new mode of urban mobility. City officials faced a choice: ban the disruptive technology or experiment with ways to integrate it. Some cities, such as San Francisco and Nashville, would put a moratorium on the vehicles, but Santa Monica's City Council sought to regulate and integrate scooters into the city's mobility ecosystem, emphasizing the potential to reduce emissions, extend transportation options in underserved areas of the city, and multiply car alternatives for navigating the crowded downtown. Over the next two years, Santa Monica contended with clutter and chaos but kept going with the scooter experiment, which eventually served thousands of riders each year. The immediate logistical and safety challenges were steep and the city relied on iterative pilot programs to learn and grow what we term their "soft infrastructure" response. This case study examines Santa Monica's experience through three innovation lenses: (1) the city as laboratory, (2) adaptive regulation, and (3) soft infrastructure.

City as Laboratory

The appearance overnight of ten swan-necked metal newcomers, followed by a couple hundred in the following days, interested and annoyed Santa Monica's leaders. This would be a typical response to the actions of Bird, a micromobility company founded by Travis VanderZanden. VanderZanden, who had worked at transportation disrupters Lyft and Uber and was no stranger to regulatory provocation, notified the city's mayor of the development over LinkedIn, writing: "We have $3M in venture funding to focus on the traffic and parking problems in Santa Monica and Venice … I'd love to work together."[30]

The company expanded at a breakneck pace, launching scooters in 120 cities over 14 months, including Los Angeles, Paris, Tel Aviv and Mexico City, and accruing half a million dollars in fines and 3 lawsuits in half that time. Drawing on the Silicon Valley playbook of "entrepreneur first and let regulators ask questions later," Bird's tactic was to find a city without laws prohibiting e-scooters, launch a fleet of them with an app that would let people find them and then ditch them anywhere, craft a pitch about climate-friendly emissions reduction and New Urbanist car-free downtowns, see how much fun people have riding them in a city with nearly 300 days of sunshine, and lay the gauntlet of what to do next at the feet of city officials.[31]

Over the next several years, many cities would ban e-scooters—either shared ones or privately owned ones, and occasionally both. But Santa Monica ultimately reacted with a lighter regulatory hand. Initially, the City implemented emergency regulations, including impound fees for parked scooters blocking right-of-ways and requirements for riders to have a valid driver's license and to wear helmets. Bird also settled with the City for operating without proper permits.[32] However, though the city had not volunteered to become a living lab, leaders also identified the

[30] Ashley Lightfoot, "The Story of Bird: The E-Mobility Brand That Flew Too Close to the Sun," *Latana*, 2022, https://resources.latana.com/post/bird-mobility-brand-deep-dive/.

[31] Jason Del Rey, "Road Kill," *Fortune*, 2024, https://fortune.com/2024/01/25/bird-scooter-bankruptcy-sharing-economy-unicorn/.

[32] Matthew Hall, "Competition Arrives for Bird Scooters," *Santa Monica Daily Press*, April 18, 2018, https://smdp.com/news/competition-arrives-for-bird-scooters/.

technology as an opportunity to experiment with how micromobility could fit urban transit needs.[33]

In a prospectus Bird released before going public, founder VanderZanden framed e-scooters as complements to existing public transportation routes. His mother had been a public bus driver during his youth in Appleton, Wisconsin, and this background, VanderZanden wrote, allowed him to imagine how e-scooters could solve the first- and last-mile challenge many public transportation riders face.[34] City leaders also received the e-scooters as part of a growing bundle of car alternatives, including the Metro E (Expo), which connects Santa Monica to Downtown Los Angeles, the Big Blue Bus system, and the Metro Bike and Breeze Bike shares (Breeze dockless bikes ceased operations in 2020),[35] all of which promised to be "cheaper, faster, healthier and/or more environmentally sustainable than owning and driving a car."[36]

In an August 2018 public memo, City Manager Rick Cole wrote, "Nobody said creating 'a new model of mobility' would be easy." But, as Cole noted, Santa Monica wanted to figure out a way to make the scooters work for the city, rather than retreat from the new mode entirely, as some neighboring cities (including West Hollywood, Beverly Hills, and Newport Beach) had done by temporarily banning them. Even in the initial weeks, the scooters, Santa Monica's deputy city manager confirmed, had proven "hugely popular" with riders even as they also "engendered a lot of concern and backlash" from residents.[37] But, city leaders wrote, "rather than ban the services, the City decided to explore, test, and evaluate shared

[33] City of Santa Monica, *Shared Mobility Pilot Report.*

[34] Bird Global, Inc., *Form S-4 Registration Statement* (U.S. Securities and Exchange Commission, 2021), https://www.sec.gov/Archives/edgar/data/1861449/000119312521162217/d431538ds4.htm.

[35] City of Santa Monica, *What's the Difference Between Breeze Bike Share and Metro Bike Share*, September 5, 2017, https://www.santamonica.gov/blog/what-s-the-difference-between-breeze-bike-share-and-metro-bike-share. Owen Gorman, *Bike Share Means Business: A Case Study of the Santa Monica Breeze Bike Share System* (University of California Los Angeles, 2016), https://ucla.app.box.com/s/3gs3kobdtzdhu9x46lh0mtc5oyev35om. Matthew Hall, "Breeze Bikeshare to End in November," *Santa Monica Daily Press*, August 14, 2020, https://smdp.com/news/breeze-bikeshare-to-end-in-november/.

[36] Rick Cole, *The Scooter Saga Continues* (City of Santa Monica, 2018), https://www.santamonica.gov/blog/the-scooter-saga-continues.

[37] Tracy Lien and Melissa Etehard, "Santa Monica Proposes Tougher Rules for Bird and Other Scooter Rental Companies," *Los Angeles Times*, June 7, 2018, https://www.latimes.com/business/technology/la-fi-tn-santa-monica-scooter-pilot-20180607-story.html.

mobility devices as a new sustainable transportation option that could potentially advance the community's broader transportation goals."[38]

"Few remember," City Manager Cole wrote, "the history of motor cars when they invaded cities and turned them upside down – or the decades of battles over enforcing safety rules and battling over how cars radically altered streets and neighborhoods. America – with the Los Angeles region in the lead – devoted trillions of dollars to convert surface transportation to cars and trucks. Millions of homes were demolished to make way for highways, tens of thousands of people died in car crashes each year and the nation became addicted to oil."

Collective memories may be short-lived, but Cole saw a generational eagerness for transportation experimentation; "older people (generally) are irate that their lives and neighborhoods are being disrupted by change," but "young people (generally) embrace the new range of options."[39] The City therefore determined to continue the experiment and moved forward with developing a competitive pilot program as a middle ground between innovation disruption and restrictive governance.

Adaptive Governance

"Advancing sustainable mobility takes planning," Santa Monica's *Shared Mobility* report began tartly; it was work the city had proven ready to do, including creating Bike Action and Pedestrian Action Plans, investing in bus and bike infrastructure and integrating local public transit with LA County's expanding light rail system. The city instituted educational and encouragement campaigns, like GoSaMo and Safe Routes to School, and launched LA County's first publicly owned bike share system. Despite this strong evidence of the city's willingness to engage in new mobility experimentation, "like many cities across the world, Santa Monica was forced to react to a new disruptive mobility model—privately owned dockless shared e-scooters and e-bikes" with no cooperative warning or planning.[40]

Historically, cities have pursued a range of governance structures to provide and regulate transportation networks for their residents. Under an ownership model, the city is directly responsible for the daily operations, staffing, and maintenance of a transit service such as Santa Monica's Big

[38] City of Santa Monica, *Shared Mobility Pilot Report.*
[39] Cole, "The Scooter Saga Continues."
[40] City of Santa Monica, *Shared Mobility Pilot Report.*

Blue Bus. Direct contracts allow cities to retain ownership but to outsource management to private companies. Santa Monica's Breeze Bike Share was one example of this structure. Another regulatory approach, conditional use permits, allows private operations that meet a general set of regulations while franchise models grant exclusive market access to one or more companies in exchange for adherence to established service standards—this is the structure Santa Monica employs to regulate local taxi services.

Yet recognizing the potential benefits of new mobility options, the city swiftly introduced a structured pilot program to channel the growing scooter experiment. City Manager Cole explained that the pilot was intended "to give the City greater control over operations" than was afforded by the "existing permit system," raise fees to cover city operational costs, and institute tougher penalties if pilot rules were disregarded.

In addition to returning clear regulatory control to the city, the pilot goals included testing oversight frameworks, aligning the new mobility service with the city's broader transportation objectives, particularly sustainability, through data collection, evaluating public-private partnership models, assessing current operators, and experimenting with innovative technologies like geofencing (the use of GPS to create a virtual boundary) to address rider behavior and scooter distribution. City Manager Cole acknowledged that bad rider behavior, especially during summer tourist months, was causing widespread frustration and demanded a regulatory response. Lime and Bird took some steps in response to local government frustration, including "tightening their software controls to discourage juveniles from riding without licenses, deploying their scooters away from the Beach Bike Path and Palisades Park and [employing a maintenance team to] promptly retriev[e] abandoned ones." The two companies also coordinated to deploy "safety ambassadors" to give away helmets and encourage riders to know the laws. But, Cole concluded, those efforts were only made in response to strong pressure from the City and did not "adequately address the widespread flouting of safety laws."[41]

Plans for the structured pilot were launched in September 2018 and drew 18 applications for scooter and bike licenses from 13 companies. Based on criteria published in the application process, five staff members ranked the applicants, with Lyft and Uber-owned Jump coming out on

[41] City of Santa Monica, *Shared Mobility Pilot Report.*

top. Insurgents Bird and Lime were conspicuous first-round losers. City Manager Cole acknowledged that some residents approvingly viewed "the pilot selection process as long overdue accountability for Bird and Lime," while others "denounced the rankings for giving the two automotive ride-sharing companies an opportunity to monopolize the emerging electric device market." The companies Bird and Lime were unsurprisingly unhappy also, protesting that they were "judged on their shortcomings while giving the benefit of the doubt to other companies who haven't actually demonstrated their ability to do any better." To demonstrate their objections, the two companies jointly shut down their scooters for a day and urged users to participate in a protest before the City Council's final decision.[42] In addition, Cole noted, City staff had received more than 10,000 emails from residents during the pilot program Public Comment period.

The final decision about who could participate in the pilot lay with Planning and Community Development Director, David Martin, who synthesized the city staff's scoring, his own evaluation, and public comments. Martin ultimately allowed four operators—Lyft, Jump, Bird, and Lime—to participate, with each company initially allowed up to 750 vehicles. Importantly, the pilot created a "performance-based fleet cap system," through which the size of each vehicle fleet could expand or contract based on how well an operator followed the city's new rules up to a maximum of 3250 vehicles total or 1 scooter for every 28 people. By comparison, Los Angeles, a city of nearly 4 million, allowed over 30,000 scooters in its initial pilot program or 1 scooter for 133 people. Permit fees would pay for additional city staff to monitor daily operations—not a small task as the city reported at the end of the pilot program having conducted weekly or monthly meetings with the four service providers, as well as creating a standardized data collection system to monitor ridership and hardware performance.[43]

Santa Monica pursued what we call an "adaptive governance" approach. The pilot set minimum requirements by ordinance and also outlined flexible expectations that could be refined through administrative regulations, which included minimum criteria on distribution and fleet size, as well as

[42] KCAL News, "Bird Scooters Protest Santa Monica Snub," *CBA News*, August 13, 2018, https://www.cbsnews.com/losangeles/news/bird-scooters-protests-santa-monica-snub/.

[43] Ryan Fonseca, "Scooters, Scooters Everywhere. Here's How LA's Grand Experiment Is Going," *LAist*, May 28, 2019, https://laist.com/news/las-big-scooter-experiment.

desired features the city hoped to prompt technology companies to pursue, including sidewalk-riding detection, systemic integration with public transit, and ADA-accessible devices. Moreover, the city proved itself willing to use the performance-capped fleet to incentivize compliance.

In July 2019, after accumulating sufficient operational data and resident complaints, the Director of Planning and Community Development reduced Bird's fleet size from 750 to 500 scooters, citing persistent noncompliance regarding proper scooter deployment, maintenance responsiveness, and inadequacies in data transparency relative to other operators.[44] The adjustment represented adaptive governance in practice, linking operational performance to regulatory consequences. The first pilot program culminated in a comprehensive report, sharing the data they had gathered on rider habits and demographics, regulatory and policy lessons learned, and persistent challenges.

During this period, the City also took stock of how the experiment had furthered their transportation goals. These were reflected in the structure of a second Shared Mobility Pilot Program. For this second pilot, the City again adopted a set of public criteria, including operator experience and past compliance, device durability, maintenance operations, safety features, education and outreach, affordability, and customer service, to select up to four operators. The Mayor stressed: "We've always taken a flexible approach to shared mobility given how quickly the industry and community needs shift."[45] Ultimately, the city selected only three operators, Spin, Veo, and Wheels, to run e-scooter and e-bike fleets during the second pilot, which ran from July 1, 2021, to September 30, 2023.[46] In the final calendar year of the project, there were over 600,000 scooter rides.[47] But, as late as 2024, the fate of scooters was still uncertain. After the conclusion of the pilot program, the City Council moved to discuss a

[44] David Martin, *Shared Mobility Pilot Program Bird Cap Fleet Size Adjustment* (City of Santa Monica, 2019).

[45] City of Santa Monica, *City Council Directs a Second Permit-Based Shared Mobility Pilot for Santa Monica*, March 10, 2021, https://www.santamonica.gov/press/2021/03/10/city-council-directs-a-second-permit-based-shared-mobility-pilot-for-santa-monica.

[46] Francie Stefan, *Shared Mobility Device Pilot Program Operator Selection and Device Allocation Memo* (City of Santa Monica, n.d.), https://www.smgov.net/uploadedFiles/Departments/PCD/Transportation/SM_Selection%20Memo_Final.pdf.

[47] Scott Snowden, "Council Moves to Establish a Permanent Shared Mobility Device Program," *Santa Monica Daily Press*, July 25, 2024, https://smdp.com/news/council-moves-ahead-to-establish-a-permanent-shared-mobility-device-program/.

permanent program, while the mayor voiced his inclination to introduce a motion to ban them entirely.[48] The outcome seemed undecided until the City Council meeting, when the Council voted to move ahead in transforming the contract model (as they had used during the pilot programs) to a permit system. Contracting with only a limited number of operators opened the system up to instability, including gaps in service and bankruptcies, which moved city planners to consider that a larger number of permitted operators might provide greater stability in service.[49] Therefore, while the Mayor continued to worry about too many scooters, the Council ultimately was concerned about too few. And, while Paris' recent ban on scooters was discussed, finally the City Council decided to expand the number of permitted operators to ensure that the scooter fleet would remain robust in the near future.

Soft Infrastructure

Bird founder Travis VanderZanden proclaimed in an interview that "Where there are no laws, that's where we go in."[50] His company supplied the technological push, and cities responded with a range of regulatory responses. From Hirschman's "hiding hand" perspective, the unannounced and uncoordinated drop of Bird scooters launched a mobility challenge that Santa Monica ultimately met, though may not have chosen for itself. Or, in the language of transportation planners, the sudden supply created its own demand, with scooters used by Santa Monica residents and visitors, and the city induced to consider what a systematic response would require.

We characterize the array of responses undertaken by Santa Monica to integrate new mobility technologies into its urban transportation ecosystem as "soft infrastructure." Unlike conventional hard infrastructure

[48] Jose Casuso, "Council Takes Up Proposal to Keep Scooters for Good," *Santa Monica Lookout*, July 22, 2024, https://www.surfsantamonica.com/ssm_site/the_lookout/news/News-2024/July-2024/07_22_2024_Council_Takes_Up_Proposal_to_Keep_Scooters_Permanently.html.

[49] Matthew Hall, "E-Scooter Debate Returns to Council yet Again," *Santa Monica Daily Press*, July 22, 2024, https://smdp.com/featured/scooter-debate-returns-to-council-yet-again/.

[50] Dara Kerr, "Bird Scooters CEO: 'Where There's No Laws, That's Where We Go In," *CNET*, October 9, 2018, https://www.cnet.com/tech/tech-industry/bird-scooters-ceo-where-theres-no-laws-thats-where-we-go-in/.

projects—large-scale investments in durable, physical structures, often requiring significant land transformation—soft infrastructure includes regulatory frameworks, informational systems, educational initiatives, and incentive structures that can largely be layered onto and integrated into the existing geographic footprint of the city. Santa Monica's soft infrastructure strategy encompassed several interconnected components: (1) modest enhancements to physical infrastructure, such as creating designated parking zones and signage for integrating scooters into existing bike lane networks; (2) robust informational infrastructure (an evolution of the cyber-physical systems of past technological regimes), built through mandatory data-sharing agreements with private providers as well as government-run user and resident surveys; (3) educational and community outreach efforts to influence rider behavior and public perceptions; (4) incentive structures to align further technological innovation with municipal goals; and (5) policies and regulatory mechanisms that could be adapted to operational feedback and emerging issues. This approach allowed the city to respond flexibly to a disruptive technology and generate valuable knowledge spillovers that have contributed to broader learning about urban micromobility governance.

Santa Monica modestly adjusted its physical infrastructure footprint to accommodate the new scooters, installing 107 designated parking and pick-up zones across the city. These zones helped organize rider parking behavior and streamlined the operations of service fleet maintenance providers. The city's introduction of e-scooters also reinforced efforts to expand and enhance its existing bike lane network so that both forms of urban mobility benefited from the relatively minor physical infrastructure changes.

Alongside these physical changes, Santa Monica invested more in its cyber and informational infrastructure systems. Data sharing was a mandatory requirement for pilot participation, and the city enforced this rigorously, as evidenced by its decision to penalize Bird when it failed to provide what the Department of Transportation felt was reliable and transparent data. The city actively solicited user feedback, conducting two user surveys (4200 completed responses) and a community-wide survey (approximately 1200 responses), that informed program evolution. Further, the city established workflows for public-private cooperation through a Community Advisory Committee that met regularly to discuss challenges and program changes. Finally, the city looked toward peer cities, hosting a Shared Micro-Mobility Summit with 15 cities, including Seattle, Portland,

Oakland, San Jose, Minneapolis, and Washington, DC, that were experimenting with their own micromobility programs.

Through this extensive informational infrastructure, Santa Monica uncovered trends in scooter usage and program challenges. The scooters proved popular, generating nearly 3 million rides between October 2018 and September 2019, with ridership peaking in spring and summer months. Trips tended to be brief, averaging 14 minutes and 1.3 miles, with popular destinations—unsurprisingly—concentrated downtown and along beach access points. This clustering created challenges for keeping scooters spread around town—critical if scooters were to become an everyday transportation option—without active rebalancing, which incurred emissions and financial costs of its own. Surveys provided significant evidence for mode shift, with nearly half of trips replacing ones that would have otherwise been made by car. However, uptake was gendered (users were over two-thirds male) and riders skewed younger and higher income. Low-income programs were available, but little utilized (~250 residents signed up), likely due to costs that were still higher than other public transit options and underinvestment in outreach and information. Finally, none of the companies deployed services for people with disabilities, such as tricycles or three-wheeled scooters.

Education and outreach campaigns were another plank in Santa Monica's soft infrastructure. The city actively encouraged, and at times pressured, scooter companies to incorporate educational components into their apps and through in-person events. These educational initiatives appeared successful, with surveys showing that 85% of riders and 90% of the general public understood basic parking and riding regulations, though awareness alone proved insufficient to eliminate ongoing issues with sidewalk riding and often obstructive parking.[51]

Having adapted physical and information infrastructure to an external technological shock, Santa Monica sought to incentivize companies to push technological innovations in a direction that suited its transportation goals. The city was among the first to enforce geofencing and digital policy tools to ameliorate parking, safety, and geographic concentration problems. The city required providers to implement a deactivation zone around the beach, which brought scooter speeds to zero within restricted areas as a direct tool for reducing safety conflicts and congestion along high-traffic pedestrian paths. While companies also introduced in-app safety features,

[51] City of Santa Monica, *Shared Mobility Pilot Report*.

the effectiveness of these measures remains uncertain, as persistent rider behavior issues suggest technological fixes alone may be insufficient without broader behavioral changes. The city also rewarded operators that improved the reliability and longevity of the technology. An analysis of early data from Louisville's Bird scooters showed that their early, consumer-grade scooters had an average lifetime of fewer than 30 days, but later generations of Bird and other operators became more durable, with fewer broken or dead scooters adding to street clutter.[52]

Finally, Santa Monica strengthened its regulatory responses by developing a structured permit and oversight system. As Bird CEO Travis VanderZanden pointed out, there were no clear rules about scooters. He argued where there were clear laws, his company did not deliberately flout them, saying, "We don't go to New York because it's technically illegal to use a scooter at the state level," but would instead target cities where the laws were vague or nonexistent.[53] In response, Santa Monica established clear rules, with city enforcement personnel issuing 299 citations covering 929 violations and impounding over 1200 scooters for violations. The city has highlighted that operators share in the responsibility not just for making reliable technology but in working with the city to develop effective socio-technical systems that may blend app-based safety information, geofenced slow-downs, and in-person outreach and safety events.

Santa Monica's experience demonstrates that the integration of new technologies—in this case a machine for individual mobility—involves iterative regulatory work. The technological shock of an unpermitted introduction of new technology—hardly likely to be the last example of this in the coming years—induced the rollout of a variety of "soft infrastructure" that Santa Monica used to buffer and accommodate the disruptive technological change. The city's Shared Mobility Reports, documenting its pilot programs, serve as the city's primary narratives for how a city can function as a space of innovation through adaptive governance. In these public documents, the city argues that rules and enforcement have contributed to more social ridership over time, but acknowledges that ongoing challenges remain, particularly with managing scooter parking, unsafe sidewalk riding, and unequal access. The case study shows that

[52] Alison Griswold, "Shared Scooters Don't Last Long," *Quartz*, March 1, 2019, https://qz.com/1561654/how-long-does-a-scooter-last-less-than-a-month-louisville-data-suggests. Seabrook, "The E-Scooters Loved by Silicon Valley."

[53] Kerr, "Bird Scooters CEO."

integration is hardly just a matter of new hardware development, but the hard-won municipal-private company detente formed around creating a workable socio-technical system that blended digital tools like app-based education and geofenced restrictions with direct outreach and educational efforts.

Travis VanderZanden's provocative admission—"Where there are no laws, that's where we go in"—captures the challenge cities face: rapid technological innovation often deliberately targets regulatory uncertainty. The follow-up question becomes: how are disruptive innovations converted into sustainable and safe aspects of a shared urban environment? Echoing Hirschman's "hiding hand" concept, the initial surprise introduction of scooters became an opportunity for the city to refine and strengthen its adaptive regulatory capacities. Though the city began by managing the disruption reactively, its regulatory innovations promise to shape the next round of technology. Moreover, the city recognized that to meet its goals around a less-crowded, lower-emission urban future, its broader transportation ecosystem needed to be disrupted and—as City Manager Rick Cole emphasized from the outset—meaningful innovations in the urban landscape involve tolerating uncertainty and navigating incremental adjustments over time.[54]

Conclusion

Micromobility companies provided an initial technological push in the form of dockless electric scooters, compelling cities to quickly generate supportive soft infrastructure—new regulations, clearly demarcated spaces for riding and parking, as well as public-private partnerships. However, the scooter itself was not the sole technological innovation. Rather, micromobility emerged from a convergence of multiple technological and economic trends. Scooter companies initially utilized consumer-grade hardware, which was inexpensive but inadequate for sustained public use. The devices broke quickly and the clutter of dead hardware only added to governance challenges. Companies pivoted to more durable hardware and better batteries. Unlike docked scooters, which initially used key fobs or credit cards for unlocking at known stations, dockless scooter systems relied on widespread smartphone use. Reliable cellphone networks, mobile app, and accurate GPS allowed customers to locate and rent the scooters.

[54] Cole, "The Scooter Saga Continues."

At the same time, shared micromobility has faced an unusual challenge with social and economic consequences since the most enthusiastic, regular, and practiced riders often choose to purchase personal vehicles. This churn is unlike enthusiastic converts to public transportation or shared e-bike programs, where assets are more out of reach for individual users, and undercuts both the customer base for shared fleets and the growth of shared norms and experience among shared scooter-system riders.[55]

These hardware and software technologies were not the only systems Silicon Valley firms have erected in the past two decades. Several micromobility companies, including early entrants like Bird, depended on a gig-economy labor force to maintain and redistribute scooters—linking micromobility to the larger labor debates and regulatory questions characteristic of platform capitalism and the sharing economy. Tensions emerged between the urbanism promises attached to micromobility innovations, such as expanding transportation access, reducing congestion, and lowering carbon emissions, and early pilot data that revealed concentrated use among higher-income users. However, recalling the multidimensional socio-technical dimensions of micromobility helps us see more clearly how the decidedly not awe-inspiring urban scooter extends Hirschman's and Kline's conceptual frameworks into the twenty-first century.

While Hirschman pointed out how ignorance of obstacles catalyzes creative problem-solving, here we might see a more deliberate *ignoring* of norms characteristic of the "move fast and break things" 2010s Silicon Valley. In Southern California, the unexpected proliferation of scooters and the ensuing social disruption forced Santa Monica into rapid, adaptive governance strategies that could not have been fully anticipated. Meanwhile, Kline's reminder that even systems technologies are ultimately locally shaped and socially constructed frames the diverse responses to scooters and puts caveats on lessons drawn from sunny, well-resourced Santa Monica. Kline's emphasis on user-driven innovation resonates strongly with the improvisational ways urban residents adapted scooters to their daily routines, creating novel transportation habits that were skewed toward certain demographics, stubbornly antisocial, but still promising enough for city operators and planners to persist, reminding their constituents that change required tolerating short-term disruption and that new technologies that helped further the city's sustainability goals deserved a pilot program, and a second, and perhaps even a third chance. As

[55] Seabrook, "The E-Scooters Loved by Silicon Valley."

micromobility users integrated scooters into urban life—whether for commuting, recreation, or last-mile connections—they contributed to a broader, evolving definition of urban mobility, raising questions about equity, access, and the reshaping of urban public spaces.

Bibliography

Beardsley, Eleanor, dir. *Paris' Rental Electric Scooter Ban Has Taken Effect*. NPR, September 1, 2023. https://www.npr.org/2023/09/01/1197329094/paris-rental-electric-scooter-ban-has-taken-effect

Berto, Frank. "Bicycle." In *Britannica*. Accessed November 16, 2025. https://www.britannica.com/technology/bicycle

Bevan, Nigel, and Miles Macleod. Usability measurement in context. *Behaviour & Information Technology* 13, nos. 1–2 (1994): 132–45. https://doi.org/10.1080/01449299408914592.

Bird Global, Inc. *Form S-4 Registration Statement*. U.S. Securities and Exchange Commission, 2021. https://www.sec.gov/Archives/edgar/data/1861449/000119312521162217/d431538ds4.htm

Boelens, Luuk. "Theorizing Practice and Practising Theory: Outlines for an Actor-Relational-Approach in Planning." *Planning Theory* 9, no. 1 (2010): 28–62. https://doi.org/10.1177/1473095209346499.

Bolton, Ogden. Electric Bicycle. U.S. Patent and Trademark Office Patent U.S. Patent No. 552,271, filed 1985.

Bureau of Transportation Statistics. "Bikeshare (Docked and Dockless) and E-Scooter Systems by Year and City Served." U.S. Department of Transportation, November 29, 2024. https://data.bts.gov/Bicycles-and-Pedestrians/Bikeshare-Docked-and-Dockless-and-E-scooter-System/cqdc-cm7d/about_data

Bureau of Transportation Statistics. *Bikeshare Ridership Exceeds Pre-Pandemic*. Bureau of Transportation Statistics, 2021. https://www.bts.gov/newsroom/bikeshare-ridership-exceeds-pre-pandemic-2019-high-fourth-straight-month

Casuso, Jose. "Council Takes Up Proposal to Keep Scooters for Good." *Santa Monica Lookout*, July 22, 2024. https://www.surfsantamonica.com/ssm_site/the_lookout/news/News-2024/July-2024/07_22_2024_Council_Takes_Up_Proposal_to_Keep_Scooters_Permanently.html

City of Santa Monica. *City Council Directs a Second Permit-Based Shared Mobility Pilot for Santa Monica*. March 10, 2021. https://www.santamonica.gov/press/2021/03/10/city-council-directs-a-second-permit-based-shared-mobility-pilot-for-santa-monica

City of Santa Monica. *Shared Mobility Pilot Program Summary Report*. 2019. https://www.smgov.net/uploadedFiles/Departments/PCD/Transportation/SantaMonicaSharedMobilityEvaluation_Final_110419.pdf

City of Santa Monica. *What's the Difference Between Breeze Bike Share and Metro Bike Share*. September 5, 2017. https://www.santamonica.gov/blog/what-s-the-difference-between-breeze-bike-share-and-metro-bike-share

Cole, Rick. *The Scooter Saga Continues*. City of Santa Monica, 2018. https://www.santamonica.gov/blog/the-scooter-saga-continues

Collingridge, David. *The Social Control of Technology*. Frances Pinter St. Martin's press, 1982.

Del Rey, Jason. "Road Kill." *Fortune*, 2024. https://fortune.com/2024/01/25/bird-scooter-bankruptcy-sharing-economy-unicorn/

DeMaio, Paul. "Bike-Sharing: History, Impacts, Models of Provision, and Future." *Journal of Public Transportation* 12, no. 4 (2009): 41–56. https://doi.org/10.5038/2375-0901.12.4.3.

Distefano, Natalia, Salvatore Leonardi, Mariusz Kieć, and Carmelo D'Agostino. 2024. "Comparison of E-Scooter and Bike Users' Behavior in Mixed Traffic." *Transportation Research Record: Journal of the Transportation Research Board*, 03611981241263339. https://doi.org/10.1177/03611981241263339.

Environmental Protection Agency (EPA), U.S. 2021. *Fast Facts: U.S. Transportation Sector GHG Emissions*. https://www.epa.gov/greenvehicles/fast-facts-transportation-greenhouse-gas-emissions

Etemad-Sajadi, Reza, Antonin Soussan, and Théo Schöpfer. "How Ethical Issues Raised by Human–Robot Interaction Can Impact the Intention to Use the Robot?" *International Journal of Social Robotics* 14, no. 4 (2022): 1103–15. https://doi.org/10.1007/s12369-021-00857-8.

European Consumer Centre Germany (ECCG). *Bike and E-Scooter Hire in Paris*. April 2024. https://www.evz.de/en/travelling-motor-vehicles/e-mobility/two-wheelers/bike-and-scooter-rental-paris.htm

Fainstein, Susan S., P. Jefferson Arminstead, and Marlene Keller, eds. *Restructuring the City: The Political Economy of Urban Redevelopment*. Longman, 1983.

Fairley, Peter. "China's Cyclists Embrace Electric Bikes - IEEE Spectrum." *IEEE Spectrum*, June 1, 2005. https://spectrum.ieee.org/chinas-cyclists-take-charge

Field, Clare, and Ihnji Jon. "E-Scooters: A New Smart Mobility Option? The Case of Brisbane, Australia." *Planning Theory & Practice* 22, no. 3 (2021): 368–96. https://doi.org/10.1080/14649357.2021.1919746.

Fonseca, Ryan. "Scooters, Scooters Everywhere. Here's How LA's Grand Experiment Is Going." *LAist*, May 28, 2019. https://laist.com/news/las-big-scooter-experiment

Geels, Frank W., and Johan Schot. "Typology of Sociotechnical Transition Pathways." *Research Policy* 36, no. 3 (2007): 399–417. https://doi.org/10.1016/j.respol.2007.01.003.

Gentner, Dedre, Brian Bowdle, Phillip Wolff, and Consuelo Boronat. "Metaphor Is like Analogy. In D. Gentner, K. J. Holyoak, & B. N. Kokinov (Eds.)." *The Analogical Mind: Perspectives from Cognitive Science*. MIT Press, 2001.

Genus, Audley, and Andy Stirling. "Collingridge and the Dilemma of Control: Towards Responsible and Accountable Innovation." *Research Policy* 47, no. 1 (2018): 61–69. https://doi.org/10.1016/j.respol.2017.09.012.

Gibson, A.E.C. 1916. Self-propelled Vehicle (U.S. Patent No. 1,192,514). U.S. Patent and Trademark Office.

Gorman, Owen. 2016. *Bike Share Means Business: A Case Study of the Santa Monica Breeze Bike Share System*. University of California Los Angeles. https://ucla.app.box.com/s/3gs3kobdtzdhu9x46lh0mtc5oyev35om

Gössling, Stefan. 2020. "Integrating E-Scooters in Urban Transportation: Problems, Policies, and the Prospect of System Change." *Transportation Research Part D: Transport and Environment* 79: 102230. https://doi.org/10.1016/j.trd.2020.102230.

Grandview Research. 2021. *Electric Scooters Market Size & Share Analysis Report, 2030*. GVR-1-68038-196-2. https://www.grandviewresearch.com/industry-analysis/electric-scooters-market

Griswold, Alison. "Shared Scooters Don't Last Long." *Quartz*, March 1, 2019. https://qz.com/1561654/how-long-does-a-scooter-last-less-than-a-month-louisville-data-suggests

Hall, Matthew. "Breeze Bikeshare to End in November." *Santa Monica Daily Press*, August 14, 2020. https://smdp.com/news/breeze-bikeshare-to-end-in-november/

Hall, Matthew. "Competition Arrives for Bird Scooters." *Santa Monica Daily Press*, April 18, 2018. https://smdp.com/news/competition-arrives-for-bird-scooters/.

Hall, Matthew. "E-Scooter Debate Returns to Council yet Again." *Santa Monica Daily Press*, July 22, 2024. https://smdp.com/featured/scooter-debate-returns-to-council-yet-again/

Heilmann, Sebastian. "Policy Experimentation in China's Economic Rise." *Studies in Comparative International Development* 43, no. 1 (2008): 1–26. https://doi.org/10.1007/s12116-007-9014-4.

Hirschman, Albert O. *Development Projects Observed*. Brookings Institution Press, 2011.

Holder, Sarah. "The Man Behind the Scooter Revolution." *Bloomberg.Com*, September 26, 2018. https://www.bloomberg.com/news/articles/2018-09-26/how-a-kids-scooter-became-a-micro-mobility-revolution

Japan Dev Team (JVT). 2024. *Electric Scooters in Japan*. https://japan-dev.com/blog/electric-scooters-in-japan

Kamen, Dean L., Robert R. Ambrogi, Robert J. Duggan, et al. Human transporter. United States Patent US5701965A, filed May 27, 1994, and issued December 30, 1997. https://patents.google.com/patent/US5701965A/en

KCAL News. "Bird Scooters Protest Santa Monica Snub." *CBA News*, August 13, 2018. https://www.cbsnews.com/losangeles/news/bird-scooters-protests-santa-monica-snub/

Kerr, Dara. "Bird Scooters CEO: 'Where There's No Laws, That's Where We Go In." *CNET*, October 9, 2018. https://www.cnet.com/tech/tech-industry/bird-scooters-ceo-where-theres-no-laws-thats-where-we-go-in/

Kline, Ronald, and Trevor Pinch. "Users as Agents of Technological Change: The Social Construction of the Automobile in the Rural United States." *Technology and Culture* 37, no. 4 (1996): 763–95.

Kline, Ronald R. *Consumers in the Country: Technology and Social Change in Rural America*. Revisiting Rural America. Johns Hopkins University Press, 2000.

Kline, Ronald R. 1992. *Steinmetz: Engineer and Socialist*. Johns Hopkins Studies in the History of Technology, new ser., no. 13. Johns Hopkins University Press.

Lakoff, George, and Mark Johnson. *Metaphors We Live by: With a New Afterword*. University of Chicago Press, 2003.

Langford, Brian Casey, Christopher Cherry, Taekwan Yoon, Stacy Worley, and David Smith. "North America's First E-Bikeshare: A Year of Experience." *Transportation Research Record: Journal of the Transportation Research Board* 2387, no. 1 (2013): 120–28. https://doi.org/10.3141/2387-14.

Lien, Tracy, and Melissa Etehard. "Santa Monica Proposes Tougher Rules for Bird and Other Scooter Rental Companies." *Los Angeles Times*, June 7, 2018. https://www.latimes.com/business/technology/la-fi-tn-santa-monica-scooter-pilot-20180607-story.html

Lightfoot, Ashley. "The Story of Bird: The E-Mobility Brand That Flew Too Close to the Sun." *Latana*, 2022. https://resources.latana.com/post/bird-mobility-brand-deep-dive/

Lindblom, Charles E. "The Science of 'Muddling Through'.'" *Public Administration Review* 19, no. 2 (1959): 79. https://doi.org/10.2307/973677.

Martin, David. 2019. *Shared Mobility Pilot Program Bird Cap Fleet Size Adjustment*. City of Santa Monica.

Moon, Christopher, Anna Bray Sharpin, Iván De La Lanza, Azra Khan, Luca Lo Re, and Anne Maassen. *The Evolution of Bike Sharing*.. January 14, 2019. https://www.wri.org/research/evolution-bike-sharing

Moreau, Hélie, Loïc De Jamblinne De Meux, Vanessa Zeller, Pierre D'Ans, Coline Ruwet, and Wouter M.J. Achten. "Dockless E-Scooter: A Green Solution for Mobility? Comparative Case Study Between Dockless E-Scooters, Displaced Transport, and Personal E-Scooters." *Sustainability* 12, no. 5 (2020): 1803. https://doi.org/10.3390/su12051803.

Mossberger, Karen, and Gerry Stoker. "The Evolution of Urban Regime Theory: The Challenge of Conceptualization." *Urban Affairs Review* 36, no. 6 (2001): 810–35. https://doi.org/10.1177/10780870122185109.

National Association of City Transportation Officials. *Shared Micromobility in the U.S.* 2020. https://nacto.org/wp-content/uploads/2020/08/2020bikesharesnapshot.pdf

NY DOT. *Shared E-Scooter Program.* 2025. https://nycdotscootershare.info/faq

Rogers, Everett M. *Diffusion of Innovations.* 5th edition. Free Press, 2003.

Schwartz, F.A.O. 1999. Opening Remarks: The States and Cities as Federal Laboratories of Democracy. *Fordham Urb. L.J.* 27. https://ir.lawnet.fordham.edu/ulj/vol27/iss1/7

Schwarz-Plaschg, Claudia. "The Power of Analogies for Imagining and Governing Emerging Technologies." *NanoEthics* 12, no. 2 (2018): 139–53. https://doi.org/10.1007/s11569-018-0315-z.

Seabrook, John. "The E-Scooters Loved by Silicon Valley Roll Into New York." *New Yorker*, April 19, 2021. https://www.newyorker.com/magazine/2021/04/26/the-e-scooters-loved-by-silicon-valley-roll-into-new-york

Shaheen, Susan A., Stacey Guzman, and Hua Zhang. "Bikesharing in Europe, the Americas, and Asia: Past, Present, and Future." *Transportation Research Record: Journal of the Transportation Research Board* 2143, no. 1 (2010): 159–67. https://doi.org/10.3141/2143-20.

Shove, Elizabeth, Mika Pantzar, and Matt Watson. *The Dynamics of Social Practice: Everyday Life and How It Changes.* SAGE Publications Ltd, 2012. https://doi.org/10.4135/9781446250655.

Snowden, Scott. "Council Moves to Establish a Permanent Shared Mobility Device Program." *Santa Monica Daily Press*, July 25, 2024. https://smdp.com/news/council-moves-ahead-to-establish-a-permanent-shared-mobility-device-program/

Stefan, Francie. *Shared Mobility Device Pilot Program Operator Selection and Device Allocation Memo.* City of Santa Monica, n.d. https://www.smgov.net/uploadedFiles/Departments/PCD/Transportation/SM_Selection%20Memo_Final.pdf

Transport for London. n.d. *Electric Scooters.* https://tfl.gov.uk/modes/driving/electric-scooter-rental-trial

Viteritti, Joseph P. "The City and the Constitution: A Historical Analysis of Institutional Evolution and Adaptation." *Journal of Urban Affairs* 12, no. 3 (1990): 221–36. https://doi.org/10.1111/j.1467-9906.1990.tb00215.x.

Von Hippel, Eric. 2005. *Democratizing Innovation.* The MIT Press. https://doi.org/10.7551/mitpress/2333.001.0001.

Yanocha, Dana, and Mackenzie Allan. 2021. *Electric Assist: Leveraging E-Bikes and E-Scooters for More Livable Cities.* Institute for Transportation & Development Policy.

Yglesias, Matthew. "The Electric Scooter Boom— and Cities' Misguided Attempts to Squash It— Explained." *Vox*, September 10, 2018. https://www.vox.com/the-goods/2018/9/10/17631318/electric-scooters-bird-city-regulations-sustainability

CHAPTER 5

Consumers at the Coastline: The Camping Tent as Deviant Technology in Twenty-First-Century Los Angeles

Christina Dunbar-Hester

For Ron, a superb mentor for caring in analysis and in life[1]

> In a swift predawn operation three days before graduation ceremonies begin, USC campus police and LAPD officers in riot gear cleared a pro-Palestinian encampment from the center of campus… There were no arrests during the sweep, according to USC officials and student activists. By about 8 a.m. some 40 tents, posters, debris and barricades had been hauled away.[2]

[1]The author wishes to thank Maggie Davis, David Beers, Eric Fielding, Chris Kelty, Cristina Visperas, and Alison Trope, among others, for sharing thoughts that shaped my thinking here; and Anto Mohsin and Jenni Lieberman for generous editorial comments that improved the chapter. All errors and omissions are of course mine.

[2]Angie Orellana Hernandez, Jaweed Kaleem, Hayley Smith, and Hannah Fry, "LAPD removes tents, clears pro-Palestinian encampment at USC," *Los Angeles Times*, May 5, 2024, https://www.latimes.com/california/story/2024-05-05/la-me-usc-camp.

C. Dunbar-Hester (✉)
University of Southern California, Los Angeles, CA, USA
e-mail: dunbarhe@usc.edu

J. L. Lieberman, A. Mohsin (eds.), *Technological Humanities*,
https://doi.org/10.1007/978-3-032-24741-4_5

2024 was the year California cracked down on homeless encampments.[3]

Introduction

In this chapter, I consider the modest polyester tent, an unremarkable, ubiquitous consumer good. I illustrate how people's deviant interpretations and uses of tents are emblematic of important facets of social geography in twenty-first-century Los Angeles (though not only here). I also trace these consumer products pre- and post-use, situating local patterns in global historical currents, in which greater Los Angeles is uniquely implicated. Following the humble camping tent reveals state-led violence and a politics of disposability.

This account follows tents as artifacts. Scanning across media narratives, marketing material, attention to tent sales and post-use, user description and interpretation, property law and urban policy, and artifactual analysis, my approach here might be called material semiotics, as described by John Law: "Material semiotics is a set of tools and sensibilities for exploring how practices in the social world are woven out of threads to form weaves that are simultaneously semiotic (because they are relational, and/or they carry meanings) and material (because they are about the physical stuff caught up and shaped in those relations)."[4] Put differently, I am interested in "how is the world in the tent, and how is the tent in the world," to borrow from Joe Dumit.[5]

Theoretically, this piece draws on Science & Technology Studies (STS) frameworks in the social construction of technology tradition, especially those emphasizing users' importance in assigning meanings to technologies.[6] It also thinks with critical geography and discard studies to visibilize patterns of labor, supply chains, and post-use that are often actively

[3] Marisa Kendall, "Encampment sweeps become widespread: 2024 California homelessness year in review," *Cal Matters*, December 30, 2024, https://calmatters.org/housing/homelessness/2024/12/california-homelessness-2024-review/.

[4] John Law, "Material Semiotics," SAGE Publications: SAGE Research Methods Foundations, 2023.

[5] Dumit, Joseph. "Writing the Implosion: Teaching the World One Thing at a Time." *Cultural Anthropology* 29, no. 2 (2014), 351.

[6] Ronald Kline and Trevor Pinch, "Users as Agents of Technological Change: The Social Construction of the Automobile in the Rural United States," *Technology and Culture* 37, no. 4 (1996): 763-795; Ronald R. Kline, *Consumers in the Country: Technology and Social Change in Rural America* (Johns Hopkins University Press, 2000); Ron Eglash, Jennifer Croissant, Giovanna Di Chiro, and Rayvon Fouché, eds., *Appropriating Technology: Vernacular Science and Social Power* (University of Minnesota Press, 2004).

obscured.[7] Los Angeles is an important global site to think with and from in order to reveal patterns of consumption and post-use.

This is a work of interpretive scholarship, which does not purport to be a systematic study. Methodologically, this chapter draws on media accounts and firsthand descriptions by tent users, shared on product websites, on social media, and with me privately as a researcher. I treat public posts on product websites as public and unlikely for harm to arise from my quoting.[8] In all other cases, I anonymize accounts I present with the aim of not having any data traced back to anyone in particular. I make this choice for two reasons: while the accounts are sociologically relevant, individuals' identities are not; and some of the activity I discuss here is controversial, stigmatized, or actively criminalized.

This chapter proceeds as follows: it establishes the historical pattern of camping tents as consumer products and then surfaces "deviant" uses of these tents. I refer to these uses as deviant because they are stigmatized and criminalized. I next turn to consider the tent in a wider arc of consumption and discard: the supply chains that bring tents to consumers and the post-use afterlives of these consumer products. I show that the deviant tent is entwined with enforced disposability under threat of state violence. Los Angeles is a particularly good (if not entirely unique) site to visibilize these patterns and consider their consequences.

Camping Out in "Nature"

Camping became a widespread recreational activity in the United States after the Civil War. Of course, camping emerged as a leisure activity only when it presented an option sufficiently distinct from everyday life. As people flocked to dwell in U.S. cities, camping provided an antidote to the lived experience of "modernization" and urbanization.[9] Visits overnight in "nature," whether on foot or accompanied by the automobile, echoed the imaginary of the frontier, which cemented for campers a social and

[7] Deborah Cowen, *The Deadly Life of Logistics: Mapping Violence in Global Trade* (University of Minnesota Press, 2014); Max Liboiron and Josh Lepawsky, *Discard Studies: Wasting, Systems, and Power* (MIT Press, 2022).

[8] Annette Markham and Elizabeth Buchanan, "Research Ethics in Context: Decision-making in Digital Research," *The Datafied Society*, ed. Karin Van Es and Mirko Tobias Schäfer (University of Amsterdam Press, 2017), 207.

[9] Terence Young, *Heading Out: A History of American Camping* (Cornell University Press, 2017).

political significance for their activities.[10] In the same period, scouting also drew people into the woods, as an organized activity meant to instill self-reliance, ruggedness, and mastery over nature. Historian Mischa Honeck argues that scouting not only "rejuvenated" society through an imaginary of a "return to nature" but promoted the formation of masculinity "in a matrix of play and adventure."[11]

Terence Young, a geographer, argues that camping's "rejection" of modernity is somewhat paradoxically underwritten by modern gear. One can only stay warm, dry, fed, and watered through reliance on technologies; and over many decades, consumer technologies especially suited to provide comfort in the midst of wilderness have made camping as a leisure activity possible. In a 1917 tract, Horace Kephart, author of many articles and books on camping, wrote that the ideal tent: "should be easy to set up. It should shed heavy rains, and should stand securely in a gale. It should keep out insects and cold draughts, but let in the rays of the campfire and plenty of pure air. It should be cool and airy on summer days, but warm and dry at night. All of which is easily said," he humorously concluded.[12] Illustrating Young's point about specialized technologies enabling consumers to rough it, Kephart recommended "a waterproof silk tent," weighing only six pounds, which, he said, "economizes cloth and weight."[13]

There is a great deal of continuity from scouting and camping as established in the late nineteenth century to today. Consumer demand for technological gear for outdoor recreation, camping, and scouting is met (and created) by major corporations devoted to marketing the outdoors and their products. Young writes, "The consumerist desire to 'smooth out' camping's adventures was soon met with 'improved' versions of tents, sleeping gear, stoves, cookware, lighting devices, and myriad other items," as early as the late nineteenth century.[14] A typical present-day example of a tent for sale is one from Recreational Equipment, Inc. (commonly

[10] Young, *Heading Out*, 10.

[11] Mischa Honeck, *Our Frontier is the World: The Boy Scouts in the Age of American Ascendancy* (Cornell University Press, 2018), 2–3.

[12] Quoted in Young, *Heading Out*, 61.

[13] Young, *Heading Out*, 61. Waterproofing silk might have involved boiling a densely woven fabric in linseed oil or coating it with paraffin (Paul Hasluck, *Cassell's Cyclopaedia of Mechanics* [Cassell and Company, Ltd., 1901]; Herbert Pearson, *Waterproofing Textile Fabrics* [The Chemical Catalog Company, 1924], 34).

[14] Young, *Heading Out*, 77.

Fig. 5.1 The advertised half-dome tent (screenshot by author, March 19, 2025)

known as REI, founded in 1938), sleeping three, listed for $400 including shipping (Fig. 5.1). The REI website lists many product specifications including its dimensions, its weight (similar to Kephart's silk tent a century ago, it is 5 pounds, 7 ounces), and its textiles (polyester taffeta for the canopy and floor; coated nylon for the rainfly). Plainly, with its price tag and features, this tent is meant for recreational camping (including backpacking with its lightweight). This purpose is underscored in a video on the product's webpage, showing the tent in situ in a woods.

Its target consumer is someone with disposable income and leisure time. Given its cost and the enumeration of its craftsmanship and features, the tent itself is meant to be relatively durable and reused over time. (The manufacturer lists enhanced durability in the fabric and includes a component to repair a tent pole should it become damaged.) One user review of this product on REI's website says: "I have been using the REI Half Dome tents for the last 20–25 years. I am looking forward to this newest model. I started using the half dome when I was in the Boy Scouts (2001–2006) and carried a half dome at Philmont Scout Ranch. I have used that tent in 32 states and 3 Countries."[15] A Los Angeles-based recreational camper who purchased a tent in 2024 told me: "I mainly [use] my 6-person tent

[15] "Half Dome 3 Tent with Footprint," REI.com, accessed March 23, 2025, https://www.rei.com/product/243736/rei-co-op-half-dome-3-tent-with-footprint.

for car camping with friends/family, and I also have a 2-person tent (my first ever tent purchase) that I use for light/short/solo camping or plan to use for backpacking."[16] This statement indicates that dedicated recreational campers might even purchase multiple tents for different purposes, all under the umbrella of leisure camping.

Appropriate(d) Technology: Deviant and Stigmatized Camping

Though the polyester tent is marketed for outdoor recreation, a frontier imaginary in an idyll, this does not mean that is its only use. Science and Technology Studies scholars have emphasized how users may use technologies in novel ways, assign different semiotic meanings to them, and even modify, hack, or creatively break them.[17] The nylon camping tent uses that interest me in this chapter do not modify the artifact itself. Rather, they put the tents to use in deviant and stigmatized ways that are unlikely to appear in marketing materials or product reviews, which came sharply into focus in 2024 in Los Angeles.

Protest Camping

The first of these examples is the protest camp. In spring 2024, students across the United States initiated Gaza solidarity protest camps on many college campuses, including the one where I am employed (Fig. 5.2). Over a series of days, students began camping out in protest of the United States' material support for Israel's war against Palestinians in Gaza that began after Hamas's attacks on Israel on October 7, 2023. In particular, students demanded that their institutions of higher education initiate an academic boycott of and financial divestment from Israel, focusing on military-industrial complex holdings in endowment investments for the latter.[18]

[16] Anonymous, personal correspondence, March 7, 2025.

[17] Kline and Pinch, "Users as Agents of Technological Change," 1996; Eglash, Croissant, Di Chiro, and Rayvon Fouché, *Appropriating Technology,* 2004.

[18] Bianca Arzán-Montañez and David Rendon, "Pro-Palestinian protesters occupy Alumni Park," *Daily Trojan,* April 25, 2024, https://dailytrojan.com/2024/04/25/pro-palestinian-protesters-occupy-alumni-park/.

Fig. 5.2 A half-dozen polyester tents with small "Free Palestine" signs affixed to them, University of Southern California campus, April 26, 2024 (photo by Cristina Visperas)

Whether or not students were consciously aware of this as they dragged colorful tents out to open spaces on their campuses, the protest camp is a prominent contemporary phenomenon and media spectacle. "Acts of camping and creating place-based sites of social reproduction and resistance can be found throughout history," becoming increasingly popular with the New Social Movements of the late 1960s, write media and geography scholars Patrick McCurdy, Anna Feigenbaum, and Fabian Frenzel.[19] Anti-apartheid protest repertoire included shantytown encampments meant to highlight the inhumane and unequal conditions of the South African apartheid regime.[20] More recently and visibly in the United States, Occupy Wall Street protest encampments arose in many cities across the country in 2011, propelling camping out as protest into public view. Young people who are college students today would have been children during Occupy; the resurgence of the protest camp in Gaza solidarity protests in 2024 points to the protest camp being a durable repertoire rather than a formation directly handed down or rehearsed from memory. Their 2024 campus encampments represent "a provisional architecture erected in opposition to formal educational architectures[, providing] a space for non-compliant learning," writes Shannon Mattern.[21]

Zeroing in on the tents themselves, it is apparent that protest camps are not an intended use of this consumer product. I did not review any marketing material naming protest camp use nor encounter any user reviews mentioning it. Nor is the protest camp tent necessarily identical to the recreational camper tent, though it may be visually quite similar. Looking for tents for sale on e-commerce giant Amazon, one can find a tent similar to the one above—in all but price, and one assumes, durability (Fig. 5.3). Priced at $12.99 or around 3% (!) of the REI tent's cost, this tent claims fairly similar features to the REI tent, including fiberglass poles and being waterproof. It is distinguished by weighing about half as much, which may be a clue as to the heft and durability of the fabric. Notably, the marketing photograph for this tent situates it in a green field near a tree—a naturalistic outdoor idyll—and the product description lists outdoor hiking, fishing, and camping as suitable uses.

[19] Patrick McCurdy, Anna Feigenbaum, and Fabian Frenzel, "Protest Camps and Repertoires of Contention," *Social Movement Studies* 15, no. 1 (2016): 97-104, 100.

[20] McCurdy, Feigenbaum, and Frenzel, "Protest Camps," 100.

[21] Shannon Mattern, "Surveillance Space: Urban infrastructures of Control," *The Architectural Review*, August 5, 2024, https://www.architectural-review.com/essays/surveillance-space-protest-encampments.

Fig. 5.3 Camouflage four-person camping dome tent (screenshot by author, March 17, 2025)

Certainly one might purchase this tent for a standard camping trip, especially if one had less disposable income than an REI shopper, if one were not a devotee of camping and only intended for it to be used for a one-off trip, or if one wished to see if one liked camping before investing in sturdier gear.

Yet in spite of the use suggested by its marketing, a very inexpensive tent available for rush delivery also holds appeal for students spontaneously deciding to set up camp on the quad of their campus. Students arriving to Gaza solidarity protest camps in spring of 2024 even prompted some observers to speculate that their activities were being coordinated and funded by nefarious outsiders: "Almost all the tents are identical—same design, same size, same fresh-out-of-the-box appearance. I know that College students are not that rich or coordinated," went a conspiratorial line of thinking.[22] This assumption belied an unfamiliarity with the repertoire of the protest camp and perhaps undue suspicion or hostility toward the students. As for the tents themselves, the far more benign and likely scenario is that students accustomed to ordering restaurant food and private car rides with the touch of a couple of buttons in a smartphone app simply procured tents the same way, enabled by frictionless e-commerce

[22] David Gilbert, "No, a Shadowy Figure Is Not Buying Tents for Columbia Student Protesters." *Wired*, April 25, 2024, https://www.wired.com/story/conspiracy-tents-student-protests-gaza/.

delivering tents to their dorm rooms.[23] They may have coordinated their purchases using social media and private messaging apps, but this does not indicate a conspiracy (see Treré on "backstage" private messaging in activism[24]).

Unhoused Neighbors

To whom else might an inexpensive and essentially disposable tent appeal? In addition to at most a few hundred college students sleeping outdoors voluntarily, tens of thousands of people in Los Angeles were without permanent, stable housing in 2024. While some slept in shelters, many lived in vehicles, and others lived outdoors in makeshift structures and tents. An official count lists around 71,000 homeless people and 4200 people sleeping in tents in L.A. County.[25]

The need for tents occurs at such a scale that they are not only purchased by individuals for individual use. And of course, people experiencing homelessness may struggle to afford to purchase a tent for shelter. Volunteer groups that refer to their activities as "mutual aid" provide direct outreach services for unhoused neighbors, including the provision of food, toiletries, clothing, and tents (see Kropotkin[26]). While the tradition of direct aid to neighbors and community members is timeless, present-day mutual aid groups in Los Angeles and around the country ramped up and systematized their activities early on in the Covid-19 pandemic. The characterization of these activities as mutual aid distinguishes them from charity services such as those provided by nonprofit organizations or the state. The website for one Los Angeles group reads, "We started this org[anization] because of the state's lack of aid & compassion during the pandemic as well as witnessing the police violence against our Black & POC [people of color] neighbors. We are fighting for liberation and freedom … We do not rely on the government or corporations.

[23] Gilbert, "No," 2024.

[24] Emiliano Treré, "The Banality of WhatsApp: On the Everyday Politics of Backstage Activism in Mexico and Spain," *First Monday* 25 (2020).

[25] Los Angeles Homelessness Services Authority, 2024 Greater Los Angeles Homeless Count, June 28, 2024, https://www.lahsa.org/news?article=976-2024-greater-los-angeles-homeless-count-data.

[26] Peter Kropotkin, *Mutual Aid: A Factor of Evolution* (McClure Phillips & Co., 1902).

Everything we do comes straight from the community, to the community."[27] Though Los Angeles already had many people living in poverty and unhoused or at the threshold of becoming unhoused, many social and economic problems became especially acute as pandemic disruptions and the virus itself led to greater health vulnerabilities, rising inflation for consumer goods including food, and spiking rents.[28] After an initial period in the early pandemic where both federal aid and state and city protections for tenants increased, eviction filings reached a decade-high for Los Angeles in June 2023.[29]

Returning to tents: one organizer with a local mutual aid group told me in early 2025 that they estimated they had purchased around 100 tents to distribute since spring 2024. (I was asking about the period that coincided with the protest camps.) Describing their mutual aid activities, they said:

> Mostly I get tents from Amazon, because they often have sales on the six person size which is the size we usually try to give people who need something to stay in long term. Cost is probably the biggest factor, and size. … it doesn't make much sense to invest in a really nice, durable tent because they are exponentially more expensive … We try to stick to the 6-person size though because one of the most important things we hear from people is that they want something they can stand up in. Everyone once in a while[,] though[,] someone has a special request for a certain kind of tent (like a pop up model or a bigger size for more people). We do our best to accommodate when we can but try not to go over about $100. The money for the tents comes almost exclusively from small donations from regular people through Venmo [a money transfer app] or [our fundraising site].[30]

This organizer's observation confirms the disparity between the middle-class leisure camping tent like the REI model and much more inexpensive tents mutual aid organizers purchase for neighbors. Somewhat to my surprise, Amazon does list "homeless" as one possible reason for tent use, in this page marketing a ten-pack of tents for $229 (or $23 apiece) (Fig. 5.4).

[27] "About us," Community Loving, accessed March 23, 2025, https://communityloving.org/about.

[28] Levi Sumagaysay, "Californians face higher costs for goods and services than before the pandemic despite inflation slowing," *Cal Matters*, March 6, 2024, https://calmatters.org/economy/2024/03/california-inflation/.

[29] Los Angeles Homelessness Services.

[30] Anonymous, personal correspondence, March 11, 2025.

Fig. 5.4 A ten-pack of tents for sale on Amazon, listing homelessness as one use (screenshot by author, March 23, 2025)

It also suggests the tents are suitable for scouting, camping, backpacking, and festivals. It is not clear whether "homeless" in the product description was added to the marketing language via human judgment or via an automated process whereby search terms users enter combine with their purchasing choices to affect product listings. In other words, this description may be downstream of search engine optimization.

Once again, comporting with the organizer's statement about durability, one can infer from the price that these tents are not especially built to last. The advertised weight confirms this too: two pounds. But the tent shown here is not a model that the mutual aid organizer would select; since it only sleeps two people, it is not roomy enough for an adult to stand in. Perusing Amazon's pages, I found another ten-pack of slightly larger (three to four people, $299, still not big enough to stand in) camping tents for sale that also listed homelessness as a "use" for the tents, but aside from these two listings (both from the same manufacturer), generally tents are sold naming camping, scouting, etc., as uses. An Amazon user review of a ten-pack said "these tents offer convenience and portability, especially for short-term or casual use," underscoring that users' expectations do not include durability.[31]

[31] "Woanger 10 Pack Bulk 2-3 Person Camping Tents for Homeless Lightweight Outdoor Tent with Rain Fly and Carrying Bag Waterproof Windproof Setup Pop Up for Homeless Camping Backpacking." Amazon.com, accessed September 21, 2025, https://www.amazon.com/Woanger-Lightweight-Waterproof-Windproof-Backpacking/dp/B0F9VYWLBX/.

Having established familiar as well as "deviant" and stigmatized uses of the camping tent, I turn toward discussion of the interconnected systems that enable both the provision of this affordable, disposable consumer product, and state violence.

The World in the Tent: Supply Chains

All of the modern tents pictured and described above are imported and made of petroleum; they are composed of modern materials, polyester or nylon (as opposed to the silk or other textiles of a century ago). We cannot understand the tent without interrogating petroleum in the local and global contexts, and the intersections where those contexts meet.

Despite its more familiar Hollywood image, Los Angeles is a port town and an oil town. Early in the twentieth century, when L.A. was far from the global city it is today, regional leaders debated where to site a major shipping port. They agreed that an expanded port was necessary to participate in growing trade enabled by and exemplified in the Panama Canal, the construction of which the United States oversaw beginning in 1904 (and under U.S. administration for the entire twentieth century before being returned to Panama in 1999), to connect its business center and its harbor to expanded international shipping. That there would be a massive port somewhere in Southern California was not, however, a matter of debate.[32] Leaders of the U.S. armed forces, fresh off armed conflict in the Philippines, also saw in Los Angeles's waterfront a means of American expansion into the Pacific realm.[33] The only debate was where to locate a major port. Eventually both Long Beach and Los Angeles answered the call, siting ports in San Pedro Bay, a notch in the coastline about twenty-five miles due south of downtown L.A.

California had been an oil-producing state prior to the industrial development of this harbor. But when regional leaders began building the ports, they were intended as traditional cargo ports. That changed very quickly with the "discovery" of oil in and near San Pedro Bay, including one of the most productive oil fields in the world in 1921, in Signal Hill (an unincorporated patch of land surrounded by Long Beach, a few miles

[32] James Tejani, "Harbor Lines: Connecting the Histories of Borderlands and Pacific Imperialism in the Making of the Port of Los Angeles, 1858–1908," *Western Historical Quarterly* 45, no. 2 (2014): 124-146.

[33] Tejani, "Harbor Lines."

inland from the ports).[34] The ports' expansion atop local oil extraction happened at a time when oil was becoming more valuable nationally, as easterners in the United States were turning to oil for heating and railroads were turning to it for fuel; in addition, automobiles were becoming more plentiful and elevating demand for fuel.[35] Yields from coastal drilling declined by the middle of the twentieth century, but the ports still participated in petroleum transshipment and do so to this day. Petroleum has been the Port of Long Beach's biggest import and export for decades, according to annual reports.[36] And Southern California is not an outlier here: crude oil, transported in tankers, constitutes around 30% of all maritime cargo globally.[37]

Both oil and the ports made modern Los Angeles, cementing it as a modern center of global capital. "The construction of the Port of Los Angeles ... defined California's and the Southwest's transition from Mexican to U.S. governance and from a heterogeneous settler society to one of Anglo and corporate dominance," writes historian James Tejani.[38] In the modern era, the Long Beach and Los Angeles ports are contiguous, though still administered separately by each city. They both still handle significant amounts of petroleum, including crude oil that enters the ports and is conveyed by pipelines to nearby refineries where it is transformed for use as gasoline and aviation fuel, and petroleum coke, a byproduct of fracking that is too dirty for domestic consumption, that is exported from Long Beach to other countries including China for consumption there. According to a Port of Long Beach spokesperson, if the port were to shut down, Southern California would run out of oil in about five days.[39]

However, both ports, and especially the Port of LA, have moved toward container cargo as an ever-greater share of revenue. Anticipating a decline in local oil production, regional managers rolled mid-century oil profits

[34] Indigenous residents of the Los Angeles Basin had long used local petroleum to seal canoes (George Cunningham and Carmela Cunningham, *Port Town: How the People of Long Beach Built, Defended and Profited from Their Harbor* [Port of Long Beach, 2016] 89), so the "discovery" of oil should not be overstated. At the same time, to identify and quantify oil reserves intended for commodification is a very different use context.

[35] Cunningham and Cunningham, *Port Town*.

[36] Christina Dunbar-Hester, *Oil Beach: How Toxic Infrastructure Threatens Life in the Ports of Los Angeles and Beyond* (University of Chicago Press, 2023).

[37] Laleh Khalili, *Sinews of War and Trade: Shipping and Capitalism in the Arabian Peninsula*, Verso Books, 2021.

[38] Tejani, "Harbor Lines," 146.

[39] Dunbar-Hester, *Oil Beach*, 12.

into port expansion. They leaned into scale in trade enabled by, among other factors, deregulation in transportation, including trucking; the creation and expansion of intermodalism and the interstate highway system, which enabled containerized freight to move along the freeways; and these ports' location on the Pacific Ocean, which was advantageous for capturing trade with China as it opened up to trade with them.[40] Containerized trade (but mainly imports) has been setting new records in San Pedro Bay most years since the 1990s. Taken together, the Port of Los Angeles and the Port of Long Beach are among the top ten container ports in the world.

This backstory situates the modern camping tent. As established above, it is made of petroleum, and the supply chains that brought it to U.S. shores also rely on cheap, abundant petroleum for overland and overseas shipping. Many tents sold in Los Angeles, and even further afield, would have entered the United States through the Port of L.A. as containerized goods shipped from Asia, probably from China. (The REI tent is listed as "imported"; the Amazon tent in Fig. 5.4 is branded Guangshuohui, indicating a Chinese company.) While the higher-priced REI tent is probably made with more durable and thus costly material, distinguishing it for the consumer, it also likely offers REI a much higher profit margin than Guangshuohui sees. It is not possible for me to compare labor costs from my vantage point, but it is possible that they are comparable if the tents were manufactured in similar locales under similar circumstances. It is also not impossible that Guangshuohui is able to sell its tent at such a low price because it is using coerced labor.[41] In this light, it is possible to view policing as situated within a global economy of coerced labor and waste-making.[42]

The Tent in the World: Post-use, Grants Pass, Reverse Logistics

The ten packs of tents and especially the Guangshuohui tent sold by Amazon are priced very affordably. They enable a consumer to click a button or two on an app and have a tent delivered on a whim, without a great deal of commitment to ownership. This might appeal to someone who

[40] Dunbar-Hester, *Oil Beach,* 63-65.

[41] See Anna Tsing, "Supply chains and the human condition," *Rethinking Marxism* 21, no. 2 (2009): 148-176 and Haleluya Hadero, "Congressional report says there's an extremely high risk Temu's supply chains have forced labor," *AP News,* June 22, 2023, https://apnews.com/article/temu-shein-forced-labor-china-de7b5398c76fda58404abc6ec5684972.

[42] Thanks to Jenni Lieberman for drawing out this insight.

knows they are going camping once but are uncertain about needing a more durable tent for repeat camping trips, or it might enable a consumer purchase that the purchaser knows is a disposable item.

In the first epigraph that opens this chapter, some forty tents from the University of Southern California students' protest camp were thrown away after a pre-dawn raid by campus and municipal police. This was the second such event in the space of ten days: L.A.P.D. had arrested over ninety people on April 24, 2024, clearing an encampment begun only that day. The *Los Angeles Times* described the tense day that culminated in the arrests: "[A]s at least two LAPD helicopters circled above throughout the afternoon, the tents at USC repeatedly went up and down, as officers with the campus Department of Public Safety told students to remove them and, at one point, dragged away lawn chairs. Students picked up their tents and walked with them in circles to avoid being in violation of a university 'no camping' policy."[43] Though the story does not describe what happened to these tents after police cleared the campus, one can assume that some were removed by protesters in retreat and others were discarded by police or custodial staff.

Throwing away polyester tents is not an anomaly in Los Angeles, even if the circumstance of the students' protest camp being forcibly tossed out of a campus is. The Los Angeles mutual aid organizer quoted above specifically referred to how authorities routinely threw out homeless people's tents in the course of "sweeps":

> The folks we work with lose tents a lot through sweeps so it doesn't make much sense to invest in a really nice, durable tent ... How long people use them really depends! Some people who are really on top of it and always are around to move their tent during sweeps might be able to keep a tent for six months or so before it starts falling apart. Others who maybe aren't as mobile or who come and go a lot will lose tents more frequently. ... By far the most frequent thing [that ends the life of a tent is] they get either thrown away or destroyed during a sweep.[44]

"Sweeping" here refers to law enforcement evicting people living in tents from the site where they are camped.[45] As the organizer explains,

[43] Angie Orellana Hernandez, Jaweed Kaleem, and Melissa Gomez, "Police arrest 93 people at USC amid Gaza war protests," *Los Angeles Times*, April 24, 2024, https://www.latimes.com/california/story/2024-04-24/usc-pro-palestinian-encampment.

[44] Anonymous, personal correspondence, March 11, 2025.

[45] Beyond the scope of this account is a discussion of what "sweeping" implies, but note that the term carries a connotation of hygienic cleaning and works to stigmatize unhoused people.

someone who anticipates a sweep and is able to preemptively move to another location might be able to stay out of harm's way and keep their tent. But if they run afoul of a sweep, people's belongings are often damaged or become lost (including not only their tents but critical belongings like family memorabilia and identification documents, thus further displacing people, both socially and legally). While it is common for sweeps to be announced in advance to warn people that they need to move along, without attractive alternatives for places to go, let alone better shelter on humane terms, this can leave people simply scrambling from one site to another, trying to stay one step ahead. Under these circumstances it makes sense for unhoused people and mutual aid organizers to favor lower-cost tents that will need to be replaced more often, as opposed to sturdier tents.

Thirty percent of homeless people in the United States live in California.[46] Los Angeles has tolerated some residents living in tents, as opposed to providing homes for them, for decades.[47] But in June 2024, the U.S. Supreme Court issued a ruling that intensified state violence against unhoused people. In its decision *City of Grants Pass v. Johnson*, colloquially referred to as *Grants Pass*, the Court held that local government ordinances that criminalize camping on public land do not constitute cruel and unusual punishment of homeless people. In the lead-up to the ruling, some cities in the United States debated whether adopting rules like the one later ratified in *Grants Pass* was a direction they wanted to take. In Neenah, Wisconsin, in March 2024, one resident said, "The people who are at their lowest do not need to be kicked out when they're down what we can't do is bully the weakest of us … Targeting these people does nothing to solve our problems and frankly it doesn't feel very Wisconsin of us."[48] In spite of many residents' fervent disapproval, Neenah passed the ordinance.

[46] Janie Har, "New study says high housing costs, low income push Californians into homelessness," *AP News*, June 20, 2023, https://apnews.com/article/homeless-california-study-poverty-high-rent-a2a4bfc9b386cb70fdd14d593f31b68c.

[47] Mike Davis, *City of Quartz: Excavating the Future in Los Angeles*, Verso Books, 2006 [1990]; see also William T. Vollman, "Homeless in Sacramento: Welcome to the new tent cities," *Harper's Magazine*, March 2011, 28-46; Chris Herring and Manuel Lutz, "The Roots and Implications of the USA's Homeless Tent Cities," *City* 19 no. 5 (2015): 689–701; Tony Sparks, *Tent City: Refusing Homelessness and Making a Home*, University of Washington Press, 2024.

[48] Paul Steeno, "Neenah passes ordinance that will impact homeless population," *WeAreGreenBay.com*, March 21, 2024, https://www.wearegreenbay.com/news/local-news/neenah-passes-ordinance-that-will-impact-homeless-population/.

Meanwhile, in California, Democrat governor Gavin Newsom leaned into the *Grants Pass* ruling with gusto, writing after its passage a state-level executive order requiring state agencies to remove homeless encampments on public property.[49] He also attracted national attention for his eager participation in "sweeping" unhoused people from their makeshift homes under a Los Angeles freeway overpass, accompanied by a camera crew.[50] As the mutual aid organizer indicates, tents—along with other important personal belongings, including identity documents and mementos—are often thrown away. Quotidian sweeps, unaccompanied by governors, reporters, and professional photographers, are a frequent occurrence, with the detritus of dwelling carted off to join the municipal waste stream.

And Los Angeles's waste stream is substantial: City of L.A. residents produce 96,500 tons of municipal solid waste daily.[51] Some of it is food, and some of it is domestically produced items. But much of this waste is downstream from the ports themselves. A glance at the ports' annual trade patterns reveals predictable patterns. The number-one trading partner is China, and much more material is imported than exported. In fact, in 2024, nearly three times more material was imported than exported.[52] According to the Port of Los Angeles, 2024 saw 1,915,678 TEU's imported (top five categories are furniture, auto parts, apparel, plastics, electronics), and 660,419 TEU's exported (top five categories are recyclable paper, pet/animal feed, soybeans, recyclable metal, automobiles).[53] (TEU stands for twenty-foot equivalent unit, i.e., the size of a standard shipping container.) Patterns are similar for Long Beach, though trade volume is higher for L.A. This reveals that top imports—among which plastics and apparel are two of the top five categories—are all items to be consumed, whereas top exports include waste/post-use material (paper

[49] Elleiana Green, "'It's mayhem and craziness': Californians react to Gavin Newsom's order to remove homeless encampments," NBC News, August 10, 2024, https://www.nbcnews.com/politics/politics-news/s-mayhem-craziness-californians-react-gavin-newsoms-order-remove-homel-rcna165401.

[50] Shawn Hubler, "Newsom Clears Homeless Camps in L.A. County, Where He Wants More 'Urgency,'" *New York Times*, August 8, 2024, https://www.nytimes.com/2024/08/08/us/newsom-homeless-los-angeles.html.

[51] "Where Does Your Los Angeles Trash Go?", AmericanReclamation.com, accessed March 24, 2025, https://www.americanreclamation.com/where-does-your-los-angeles-trash-go/.

[52] Annual Facts and Figures Card, Port of Los Angeles, 2024, accessed March 24, 2025, https://www.portoflosangeles.org/business/statistics/facts-and-figures.

[53] This is just containerized imports: it does not include imports that are handled differently, ranging from liquid petroleum to automobiles and much in between.

and metal). This pattern of consumption raises questions about consumption's afterlives.

After students' and homeless people's tents have outlived their utility, through wear and tear or through state violence, the tents (and all manner of other solid municipal waste, as well as industrial waste, etc.) fan out to area landfills in Los Angeles County hinterlands.[54] The city of Long Beach operated a trash incinerator near the Port of Long Beach for decades, burning hundreds of tons of trash daily for over thirty-five years, generating electricity used in port operations and emitting toxic emissions, but it shuttered in 2024 due to emissions concerns.[55] Some textile waste can be reclaimed and "upcycled" rather than landfilled, but this is only viable for cotton, not petrochemical products like nylon and polyester.[56] Polyester and nylon are bound for landfills, where they may degrade into microplastics. Tents are certainly only a minuscule fraction of landfill waste, but they are illustrative of bigger patterns.

One of these is that once consumer items are purchased, the struggle to keep them from becoming trash ensues immediately. E-commerce enables blissfully easy consumption, but if a customer decides against keeping a product, it enters a stream of goods management that has been dubbed "reverse logistics."[57] And indeed, many more goods purchased online get returned than those purchased in person; around 30% of online purchases are returned as opposed to fewer than 10% of brick-and-mortar store purchases.[58] Consumer goods that are sold very cheaply do not price in the space or labor to deal with sorting and reselling returned items, even though many returned items are not defective. Retailers contract other

[54] Brian Taylor, "Landfill expansion rejected over groundwater concerns," *Waste Today*, March 12, 2024, https://www.wastetodaymagazine.com/news/waste-connections-castaic-la-california-landfill-expansion-permit-denial-groundwater/; "Where Does Your Los Angeles Trash Go?," AmericanReclamation.com.

[55] Jason Ruiz, "Long Beach to close its trash-burning facility, making garbage collection more expensive," *Long Beach Post*, January 23, 2024, https://lbpost.com/news/long-beach-to-close-its-trash-burning-facility-making-garbage-collection-more-expensive/.

[56] Shelley Mann, "Los Angeles textile recovery pilot diverts 1,000 pounds of waste," *Waste Today*, October 2, 2024, https://www.wastetodaymagazine.com/news/los-angeles-textile-recovery-pilot-diverts-1000-pounds-of-waste/.

[57] Harriet Constable, "Your brand new returns end up in landfill," *BBC Earth*, accessed March 25, 2025, https://www.bbcearth.com/news/your-brand-new-returns-end-up-in-landfill.

[58] "Amazon Return Pallets: How and Where to Buy Them," DirectLiquidation.com, accessed March 24, 2025. https://www.directliquidation.com/liquidation-102/buy-amazon-customer-returns-pallets-online/.

companies that ship the returned goods elsewhere for resale or simply discard them.[59]

Amazon is the behemoth of e-commerce, with millions of daily sales. It handles many returns by sorting the goods into bundled pallets and selling those to third-party buyers, who are then tasked with trying to turn a profit on the mystery goods they receive in the pallet. I reviewed a video on the streaming platform YouTube of someone who purchased an Amazon pallet and was evaluating the resale value of the items and strategizing about how to turn, let alone maximize a profit from them. To my surprise, the pallet contained multiple tents.[60] I do not have reason to believe that tents are especially meaningful as a percentage of overall consumer goods floating along in reverse logistics streams, but it was not clear that the pallet purchaser was going to be able to resell or even appropriately donate many of the goods. The tents can thus stand in for and draw attention to a much more vast stream of cheap consumer products, many made of plastic. They are but a drop in raging rivers of trash-in-formation enabled by ready access to petroleum as raw material for goods fabrication and fuel.

Conclusion: Disposability and Violence

The cheap polyester tent might be a paradigmatic technology of twenty-first-century Los Angeles, hiding in plain sight. In spite of its marketing for outdoor leisure and recreation, some meaningful number of tent purchases are actually for deviant consumption, including shelter for unhoused people and (more infrequently) protest camps. Cheap tents—whether purchased for one-off recreational camping trips, for living on public land, or for protest camps, and whether they are used up by the consumer, thrown out by police, or returned to the retailer—are all destined for the landfill, often rather quickly. Tents reveal disposability, violence, and precarity in both Los Angeles and in the world systems that bring cheap tents to their consumers.

Cheap, thin fabric that won't hold up to reuse or will be violently thrown away before it is used up stands in for a much bigger system of

[59] Constable, "Your brand new."

[60] Danny Gonzalez, "Unboxing Abandoned Amazon Returns," December 11, 2024, https://www.youtube.com/watch?v=mBwEGPXd_yg.

mistreatment of fossil neighbors and of violent disposability.[61] Scholars working in discard studies remind us that we should "trouble the assumptions, premises, and popular mythologies of waste so discussions can address wider systems and power dynamics rather than remain mired in technological or moral fixes … that deal with symptoms rather than origins of problems."[62] In Los Angeles, the histories of oil and of offshored labor, supply chains stretching across oceans and continents, provide clues to some of the problems here.[63] Indeed, as geographer Deborah Cowen points out, supply chains themselves are secured by violence and designed to restrict human movement while enabling a flow of goods: "Protecting trade networks from disruption creates new spaces of security and in doing so problematizes the political and legal status of subjects."[64] Unpalatable and even coerced work conditions for workers in far locales connect them to people that states deem criminal or disposable closer to the site of consumption.[65]

The tent also provides a way into thinking about logics of land, property, and dispossession that operate here. Needless to say, charting the frontier itself and its later rehearsal in recreational camping are settler practices. STS scholar Nicholas Shapiro writes, "Settler colonialism can be thought of as a process of manufacturing home on top of and in place of the homes of a land's original human inhabitants."[66] Shapiro is writing about the mobile home, a more durable structure than a tent. But mobile homes and tents both transmuted from being structures that initially helped settlers cover and claim more ground into where poorer and often racialized people now dwell after being shut out of more durable middle-class homes. They are architectures of dispossession in more ways than one.

California had in 2024 around 180,000 unhoused people, about one-third of the national total.[67] While California residents might argue for or against the *Grants Pass* decision, it is difficult to imagine anyone in

[61] Zoe Todd, "Fish, Kin and Hope: Tending to Water Violations in Amiskwaciwâskahikan and Treaty Six Territory," *Afterall: A Journal of Art, Context and Enquiry* 43, no. 1 (2017): 102-107.

[62] Liboiron and Lepawsky, *Discard Studies*, 6.

[63] Dunbar-Hester, *Oil Beach*; Cowen, *Deadly Life*.

[64] Cowen, *Deadly Life*, 12.

[65] See also Tsing, "Supply chains."

[66] Nicholas Shapiro, "Manufacturing Home," *Journal for the Anthropology of North America* 22, no. 2 (2019): 121-124, 121.

[67] Green, "'It's mayhem and craziness.'"

California arguing that to treat unhoused neighbors cruelly "is not who we are," which was at least subject to debate in Wisconsin.[68] California is home to just over a tenth of the U.S. population, so its homelessness rate is disproportionate. Indeed, the normalization of state violence against unhoused people in California is so great that even during massive, historic Los Angeles wildfires in January 2025, police still turned out to "sweep" unhoused people from the streets in other parts of the city even while fires actively burned and ash rained down.[69] Not least because California wildfires have led to more people becoming displaced and homeless, this heaps irony upon cruelty.[70]

In this regard, the deviant tent consumer and the dominant one are potentially entangled. The Angeleno who above described buying a tent for recreational camping also told me: "I have collected camping gear after this tent purchase not only to make my camping experience better, but to also be prepared for any possible weather emergencies (i.e environmental, social, political, etc). Inspiration for this preparation are also the experiences of refugees and displaced Palestinians."[71] In 2021, the Federal Emergency Management Agency (FEMA) published its first National Risk Index, which ranked Los Angeles County number one in a list of vulnerability to disasters. The ranking takes into account economic loss from natural disasters, and "social vulnerability," defined as "the susceptibility of social groups to the adverse impacts of natural hazards, including disproportionate death, injury, loss or disruption of livelihood."[72] Many Angelenos live with a feeling of unease due to elevated risk of fires and earthquakes but also due to the visible and palpable state abandonment experienced by neighbors who sleep in tents along riverbanks and under

[68] See Richard Barbrook and Andy Cameron, "The Californian Ideology," *Science as Culture* 6, no. 1 (1996): 44-72.

[69] I did not see this confirmed in mainstream news media; only shared on social media (January 9, 2025). However, the social media reports I saw were from mutual aid organizers I trust to report this honestly, and news media was preoccupied.

[70] Marisa Kendall, "How the LA fires could exacerbate California's homelessness crisis," *Cal Matters*, January 15, 2025, https://calmatters.org/housing/homelessness/2025/01/la-fires-homeless/.

[71] Anonymous personal correspondence, March 7, 2025.

[72] "FEMA ranks LA County as most vulnerable in nation for natural disasters," *Long Beach Post*, January 3, 2021, https://lbpost.com/news/fema-ranks-la-county-as-most-vulnerable-in-nation-for-natural-disasters/.

overpasses.[73] Students also made this connection: in their list of demands directed at the university, they included "Stop the Displacement, from South Central [gentrification, some of which is turbocharged by USC's real estate investments] to Palestine."[74]

The deviant tent is a technology of simultaneously modest and grand ambition. As mutual aid organizers have underscored, it aims to improve lives while professing little faith in states or traditional modernist infrastructure,[75] but it also embodies solidarities among community members preparing to support one another through whatever social, economic, and "natural" disasters are yet to unfold. For students in a protest camp, tents mark the creation of a space for reproducing daily life (if a more prefigurative version of it), as well as round-the-clock protest that takes aim at the hardened target that is the foreign policy of the U.S. empire. Both forms invite violent retribution by the state—a state which itself rests on a violent history (and present) of camping on land in order to lay (and rehearse) claim to it.

Thus, the cheap polyester tent is ultimately a flimsy artifact sitting atop very deep and longstanding conflicts over land, power, waste, and property.[76] The state's insistence that protesters and homeless people desist from camping on land that the state itself had expropriated from prior inhabitants might be an amusingly ironic contradiction if it were not accompanied by so much violence. Still, the deviant tent, by unsettling the familiar tent, may point to ways to imagine less toxic and more durable dwelling practices, including land restitution and more convivial relations among inhabitants.

[73] This account does not have space to account for nuances such as the historical hobo (Todd DePastino, *Citizen Hobo: How a Century of Homelessness Shaped America*, University of Chicago Press, 2010).

[74] Jonathan Park, "What protesters are demanding as they occupy Alumni Park," *Daily Trojan*, April 24, 2024, https://dailytrojan.com/2024/04/24/what-protesters-are-demanding-as-they-occupy-alumni-park/.

[75] Peter Redfield, "Fluid Technologies: The Bush Pump, the LifeStraw® and Microworlds of Humanitarian Design," *Social Studies of Science* 46, no. 2 (2016): 159-183.

[76] Shapiro, "Manufacturing Home, 123." See also Kyle Powys Whyte, "Is It Colonial Déjà Vu? Indigenous Peoples and Climate Injustice," in *Humanities for the Environment: Integrating Knowledges, Forging New Constellations of Practice*, edited by Joni Adamson, Michael Davis, and Hsinya Huang (Routledge/Earthscan Publications, 2017), 88–104.

Bibliography

Amazon.com. "Woanger 10 Pack Bulk 2-3 Person Camping Tents for Homeless Lightweight Outdoor Tent with Rain Fly and Carrying Bag Waterproof Windproof Setup Pop Up for Homeless Camping Backpacking." Accessed September 21, 2025. https://www.amazon.com/Woanger-Lightweight-Waterproof-Windproof-Backpacking/dp/B0F9VYWLBX/

AmericanReclamation.com. "Where Does Your Los Angeles Trash Go?" Accessed March 24, 2025. https://www.americanreclamation.com/where-does-your-los-angeles-trash-go/

Arzán-Montañez, Bianca and David Rendon. "Pro-Palestinian protesters occupy Alumni Park." *Daily Trojan,* April 25, 2024. https://dailytrojan.com/2024/04/25/pro-palestinian-protesters-occupy-alumni-park/

Barbrook, Richard, and Andy Cameron. The Californian Ideology. *Science as Culture* 6, no. 1 (1996): 44-72.

Community Loving. "About us." Accessed March 23, 2025. https://communityloving.org/about

Constable, Harriet. "Your brand new returns end up in landfill." *BBC Earth.* Accessed March 25, 2025. https://www.bbcearth.com/news/your-brand-new-returns-end-up-in-landfill

Cowen, Deborah. *The Deadly Life of Logistics: Mapping Violence in Global Trade.* University of Minnesota Press, 2014.

Cunningham, George, and Carmela Cunningham. *Port Town: How the People of Long Beach Built, Defended and Profited from Their Harbor.* Port of Long Beach, 2016.

Davis, Mike. *City of Quartz: Excavating the Future in Los Angeles.* Verso Books, 2006 [1990].

DePastino, Todd. *Citizen Hobo: How a Century of Homelessness Shaped America.* University of Chicago Press, 2010.

DirectLiquidation.com. "Amazon Return Pallets: How and Where to Buy Them." n.d., accessed March 24, 2025. https://www.directliquidation.com/liquidation-102/buy-amazon-customer-returns-pallets-online/

Dumit, Joseph. "Writing the Implosion: Teaching the World One Thing at a Time." *Cultural Anthropology* 29, no. 2 (2014): 344-362.

Dunbar-Hester, Christina. *Oil Beach: How Toxic Infrastructure Threatens Life in the Ports of Los Angeles and Beyond.* University of Chicago Press, 2023.

Eglash, Ron, Jennifer Croissant, Giovanna Di Chiro, and Rayvon Fouché, *Appropriating Technology: Vernacular Science and Social Power.* University of Minnesota Press, 2004.

Gilbert, David. "No, a Shadowy Figure Is Not Buying Tents for Columbia Student Protesters." *Wired*, April 25, 2024. https://www.wired.com/story/conspiracy-tents-student-protests-gaza/

Gonzalez, Danny. "Unboxing Abandoned Amazon Returns." December 11, 2024. https://www.youtube.com/watch?v=mBwEGPXd_yg

Green, Elleiana. "'It's mayhem and craziness': Californians react to Gavin Newsom's order to remove homeless encampments." *NBC News*, August 10, 2024. https://www.nbcnews.com/politics/politics-news/s-mayhem-craziness-californians-react-gavin-newsoms-order-remove-homel-rcna165401

Hadero, Haleluya. "Congressional report says there's an extremely high risk Temu's supply chains have forced labor." *AP News*, June 22, 2023. https://apnews.com/article/temu-shein-forced-labor-china-de7b5398c76fda58404abc6ec5684972

Har, Janie. "New study says high housing costs, low income push Californians into homelessness." *AP News*, June 20, 2023. https://apnews.com/article/homeless-california-study-poverty-high-rent-a2a4bfc9b386cb70fdd14d593f31b68c

Hasluck, Paul. *Cassell's Cyclopaedia of Mechanics*. Cassell and Company, Ltd., 1901.

Herring, Chris, and Manuel Lutz. "The Roots and Implications of the USA's Homeless Tent Cities." *City* 19, no. 5 (2015): 689–701.

Honeck, Mischa. *Our Frontier Is the World: The Boy Scouts in the Age of American Ascendancy*. Cornell University Press, 2018.

Hubler, Shawn. "Newsom Clears Homeless Camps in L.A. County, Where He Wants More 'Urgency.'" *New York Times*, August 8, 2024. https://www.nytimes.com/2024/08/08/us/newsom-homeless-los-angeles.html

Kendall, Marisa. "Encampment sweeps become widespread: 2024 California homelessness year in review." *Cal Matters*, December 30, 2024. https://calmatters.org/housing/homelessness/2024/12/california-homelessness-2024-review/

Kendall, Marisa. "How the LA fires could exacerbate California's homelessness crisis." *Cal Matters*, January 15, 2025. https://calmatters.org/housing/homelessness/2025/01/la-fires-homeless/

Khalili, Laleh. *Sinews of War and Trade: Shipping and Capitalism in the Arabian Peninsula*. Verso Books, 2021.

Kline, Ronald R. *Consumers in the Country: Technology and Social Change in Rural America*. Johns Hopkins University Press, 2000.

Kline, Ronald, and Trevor Pinch. "Users as Agents of Technological Change: The Social Construction of the Automobile in the Rural United States." *Technology and Culture* 37, no. 4 (1996): 763-795.

Kropotkin, Peter. *Mutual Aid: A Factor of Evolution*. McClure Phillips & Co., 1902.

Law, John. "Material Semiotics." SAGE Publications: SAGE Research Methods Foundations, 2023.

Liboiron, Max and Josh Lepawsky. *Discard Studies: Wasting, Systems, and Power*. MIT Press, 2022.

Long Beach Post. "FEMA ranks LA County as most vulnerable in nation for natural disasters," January 3, 2021. https://lbpost.com/news/fema-ranks-la-county-as-most-vulnerable-in-nation-for-natural-disasters/

Los Angeles Homelessness Services Authority. 2024 Greater Los Angeles Homeless Count. June 28, 2024. https://www.lahsa.org/news?article=976-2024-greater-los-angeles-homeless-count-data

Mann, Shelley. "Los Angeles textile recovery pilot diverts 1,000 pounds of waste." *Waste Today*, October 2, 2024. https://www.wastetodaymagazine.com/news/los-angeles-textile-recovery-pilot-diverts-1000-pounds-of-waste/

Markham, Annette, and Elizabeth Buchanan. "Research Ethics in Context: Decision-Making in Digital Research." In *The Datafied Society*, edited by Karin Van Es and Mirko Tobias Schäfer. University of Amsterdam Press, 2017.

Mattern, Shannon. "Surveillance Space: Urban Infrastructures of Control." *The Architectural Review*, August 5, 2024. https://www.architectural-review.com/essays/surveillance-space-protest-encampments

McCurdy, Patrick, Anna Feigenbaum, and Fabian Frenzel. "Protest Camps and Repertoires of Contention." *Social Movement Studies* 15, no. 1 (2016): 97-104.

Orellana Hernandez, Angie, Jaweed Kaleem, and Melissa Gomez. "Police arrest 93 people at USC amid Gaza war protests." *Los Angeles Times*, April 24, 2024. https://www.latimes.com/california/story/2024-04-24/usc-pro-palestinian-encampment

Orellana Hernandez, Angie, Jaweed Kaleem, Hayley Smith, and Hannah Fry. "LAPD removes tents, clears pro-Palestinian encampment at USC." *Los Angeles Times*, May 5, 2024. https://www.latimes.com/california/story/2024-05-05/la-me-usc-camp

Park, Jonathan. "What protesters are demanding as they occupy Alumni Park." *Daily Trojan*, April 24, 2024. https://dailytrojan.com/2024/04/24/what-protesters-are-demanding-as-they-occupy-alumni-park/

Pearson, Herbert. *Waterproofing Textile Fabrics*. The Chemical Catalog Company, 1924.

Port of Los Angeles. Annual Facts and Figures Card, 2024. Accessed March 24, 2025. https://www.portoflosangeles.org/business/statistics/facts-and-figures

Redfield, Peter. "Fluid Technologies: The Bush Pump, the LifeStraw® and Microworlds of Humanitarian Design." *Social Studies of Science* 46, no. 2 (2016): 159-183.

REI.com. "Half Dome 3 Tent with Footprint." n.d., accessed March 23, 2025. https://www.rei.com/product/243736/rei-co-op-half-dome-3-tent-with-footprint

Ruiz, Jason. "Long Beach to close its trash-burning facility, making garbage collection more expensive." *Long Beach Post*, January 23, 2024.

https://lbpost.com/news/long-beach-to-close-its-trash-burning-facility-making-garbage-collection-more-expensive/
Sparks, Tony. *Tent City: Refusing Homelessness and Making a Home*. University of Washington Press, 2024.
Steeno, Paul. "Neenah passes ordinance that will impact homeless population." *WeAreGreenBay.com*, March 21, 2024. https://www.wearegreenbay.com/news/local-news/neenah-passes-ordinance-that-will-impact-homeless-population/
Sumagaysay, Levi. "Californians face higher costs for goods and services than before the pandemic despite inflation slowing." *Cal Matters*, March 6, 2024. https://calmatters.org/economy/2024/03/california-inflation/
Taylor, Brian. "Landfill expansion rejected over groundwater concerns." *Waste Today,* March 12, 2024. https://www.wastetodaymagazine.com/news/waste-connections-castaic-la-california-landfill-expansion-permit-denial-groundwater/
Tejani, James. "Harbor Lines: Connecting the Histories of Borderlands and Pacific Imperialism in the Making of the Port of Los Angeles, 1858–1908." *Western Historical Quarterly* 45, no. 2 (2014): 124-146.
Todd, Zoe. "Fish, Kin and Hope: Tending to Water Violations in Amiskwaciwâskahikan and Treaty Six Territory." *Afterall: A Journal of Art, Context and Enquiry* 43, no. 1 (2017): 102-107.
Treré, Emiliano. "The Banality of WhatsApp: On the Everyday Politics of Backstage Activism in Mexico and Spain." *First Monday* 25 (2020).
Tsing, Anna. "Supply Chains and the Human Condition." *Rethinking Marxism* 21, no. 2 (2009): 148-176.
Vollman, William T. 2011. Homeless in Sacramento: Welcome to the New Tent Cities. *Harper's Magazine*, March 2011.
Whyte, Kyle Powys. "Is It Colonial Déjà Vu? Indigenous Peoples and Climate Injustice." In *Humanities for the Environment: Integrating Knowledges, Forging New Constellations of Practice*, edited by Joni Adamson, Michael Davis, and Hsinya Huang. Routledge/Earthscan Publications, 2017.
Young, Terence. *Heading Out: A History of American Camping*. Cornell University Press, 2017.

CHAPTER 6

Developing Dispossession: Retrofitting Settler Colonialism at the Huntly Power Station

Nathan Kapoor and Suzanne Moon

Introduction

In the early 1970s, the town of Huntly in Aotearoa-New Zealand campaigned to become the site of a major, coal-fired thermoelectric plant. Advocates saw the plant as an opportunity to revive the declining mining town's fortunes by supplying electricity more reliably than chancy hydroelectric installations could.[1] Narratives of rescue like those asserted by

[1] Māori language terms commonly used in New Zealand appear throughout this chapter without italics. Aotearoa-New Zealand, New Zealand, and Aotearoa are all names that the people there use to reference the islands. In Māori, Aotearoa means "land of the long white

N. Kapoor
Illinois State University, Normal, IL, USA

S. Moon (✉)
University of Oklahoma, Norman, OK, USA
e-mail: suzannemoon@ou.edu

J. L. Lieberman, A. Mohsin (eds.), *Technological Humanities*,
https://doi.org/10.1007/978-3-032-24741-4_6

Huntly's advocates often figure prominently in infrastructure projects. In modernization and development efforts and anti-poverty programs, electric power is portrayed as a vital tool for recovering places and people left out of the promises of modern economies.

Yet taking the notion of rescue too much at face value runs the risk of obscuring the contexts in which infrastructures operate and the purposes that infrastructures serve in political and social life. If we interrogate infrastructures not as just things but also as relationships, the character of the relationships fostered by infrastructure projects offers insight into the broader sociopolitical meanings of technological work.[2] We are especially interested in the spatial and temporal aspects of these relationships. Electricity infrastructure projects often function to structure or restructure relationships between places, including in this story the relationship between rural and urban areas, and define the role each plays with respect to the other in supporting larger national ideals.[3] In Aotearoa-New Zealand, rural-urban relationships are inextricably bound to a history of Māori dispossession and settler utopianism. Indeed, the binary of rural-urban emerges from settler ways of understanding place; Māori forms of place-making and place-understanding are largely incommensurate with settler thinking. Electrification projects thus provoke tension as much as hope in Aotearoa-New Zealand's wider society.

Yet understandings of place are also intertwined with temporality. Howe et al. remind us that the siting, design, redesign, and use of infrastructures can narrate particular stories about the relationship between the

cloud." New Zealand comes from the Dutch name, Nova Zeelandia, named for Zeeland; For more on the Huntly Power Station and the region, see Jo Whittle, "Into the Backyard: Huntly Power Station and the History of Environmentalism in New Zealand," *Environment and Nature in New Zealand* 8 no 1 (2013): 7–38 and Marama Muru-Lanning, *Tupuna Awa: People and Politics of the Waikato River* (Auckland University Press, 2016), 63–64.

[2] Brian Larkin, "The Politics and Poetics of Infrastructure," *Annual Review of Anthropology* 42 (2013): 327–343.

[3] However, the national scale—politically and ideologically—is not always the most relevant to examine. Sunila Kale, "Structures of Power: Electrification in Colonial India," *Comparative Studies of South Asia, Africa and the Middle East*, 34, no 3 (2014): 454–475; Casey P. Cater, *Regenerating Dixie: Electric Energy and the Modern South* (Pittsburgh, University of Pittsburgh Press. 2019). On sociotechnical imaginaries, see Sheila Jasanoff and Sang-hyun Kim, *Dreamscapes of Modernity: Sociotechnical Imaginaries and Fabrications of Power* (Chicago: University of Chicago Press, 2015).

past, present, and future.[4] Their concept of retooling/retrofit highlights how change to infrastructures can "bridge timelines – from past to the present and from the present to the future."[5] Going beyond claims regarding rescue for either Huntly or Aotearoa-New Zealand, the history of the Huntly power station project is best understood by its intersecting narratives about the desired relationship between urban and rural spaces, between Pākehā and Māori spaces, and between the past, present, and future in the story of the nation. [6] In Aotearoa-New Zealand, settler-colonial traditions and priorities dominated the construction of the Huntly power station but not without opposition from those who envisioned infrastructure that could tell different stories about the nation's past and future.

Our analysis was inspired by Ron Kline's *Consumers in the Country*, especially its attention to the anxieties surrounding rural-urban relationships that spurred calls for electrification of the countryside in the early twentieth-century United States.[7] Superficially, this too was a story of "rescue" as the Country Life Commission, convened by Theodore Roosevelt, responded to worries that unrelieved drudgery and isolation, and the lure of city life was causing people to abandon rural areas, undermining the settler-agricultural foundation of American life. The commission asserted that new technologies, including electricity, would make life in the country as convenient and satisfying as it was in the city. Yet farm communities saw their problems rather differently; Kline demonstrated the agency of agricultural people in shaping the direction of rural technological change based on their visions of a desirable future. Using a concept that emerged after the publication of *Consumers in the Country*, Kline offers insight into two overlapping, but not identical, sociotechnical imaginaries.[8]

Reflecting on Kline's exploration of rural electrification through the concept of the sociotechnical imaginary, we found interesting parallels with the Huntly case regarding how space and time matter in sociotechnical imaginaries. In both histories, defining the relationship between rural

[4] Cymene Howe et al. "Paradoxical Infrastructures: Ruins, Retrofit, and Risk," *Science, Technology, & Human Values* 41, no. 3 (May 2016):547–565.

[5] Howe et al., "Paradoxical Infrastructures."

[6] Pākehā refer to people of non-Polynesian descent in Aotearoa-New Zealand.

[7] Ronald Kline, *Consumers in the Country: Technology and Social Change in Rural America* (Baltimore: Johns Hopkins University Press, 2002).

[8] Jasanoff and Kim, *Dreamscapes of Modernity*.

and urban places and peoples held moral, political, and economic urgency, and the configuration of those relationships revealed how new technologies were intended to connect the past, present, and future.[9] For example, the Country Life Commission's focus on farmers and farm communities, not other rural peoples or places, betrayed their particular concern with farmers' place in American society. By blaming farm migration on local hardships and the lure of implicitly immoral city life, rather than on the serious problems posed by settler agriculture, especially in the American West, the Commission portrayed rural electrification as a moral and political project as much as an economic one.[10] The ideal of America as a land of yeoman settler-farmers was at stake. New electrical infrastructure was a way of retooling the larger sociotechnical organization of rural life in the service of stabilizing America's settler-colonial ideals. The normative question—what should America be?—was as much built into the project as the question of how farm life should change.

By likewise conceptualizing changes to Aotearoa-New Zealand's entire system of energy supply as a form of retooling, we show how shifting imaginaries linking rural and urban were simultaneously bridging timelines, linking the past to the present in normative and politically significant ways. Breaking with an earlier history of electrification reliant on hydropower, the Huntly power station would be thermoelectric, running on the output of Huntly's coal mines or natural gas drilled in the North Island. Advocates saw this retooling as a way to meet changing aspirations for energy security in the post-World War II period; promoting the growth of Auckland; reviving the fortunes of Huntly, a small, rural, coal-mining town; and securing Aotearoa-New Zealand's autonomy in a rapidly changing geopolitical environment.

Yet Huntly was also a place of power and meaning for the Waikato-Tainui (the most populous Māori iwi or tribe in the Waikato, where Huntly resides), and, during the time of the power station's construction, the epicenter of emerging environmental concerns.[11] For Māori and their supporters, their long history of dispossession at the hands of settlers

[9] Howe et al., "Paradoxical Infrastructures."

[10] Donald Pisani, "Federal Reclamation in the American West in the Twentieth Century," *Agricultural History* 77, no. 3: 391–419.

[11] The Waikato-Tainui are a collection of iwi (tribes) of the Māori and one of the most prominent in Waikato. Marama Muru-Lanning, "The Analogous Boundaries of Ngaati Mahuta, Waikato-Tainui, and Kiinitanga" *The Journal of the Polynesian Society*, 120, no. 1: 9–41. http://www.jstor.org/stable/23041477.

required as much attention and correction as energy security. Their desired trajectory from the past to the future envisioned a power station built elsewhere and restored Māori control of places of significance. Thus, different imaginaries of the trajectory linking Aotearoa-New Zealand's past and future were at play in the Huntly project. The Huntly station, as promoted by public works officials, tied modern energy production to the same settler-colonial priorities that had dominated Aotearoa-New Zealand in the past, reconfiguring the role of urban and rural in national ideals without challenging fundamental thinking about who should have the final say about the use of land and the good of the nation.

The Rural and the Urban in Aotearoa-New Zealand

In 2025, Auckland, with a population of approximately 1.5 million people, accounts for about 33% of Aotearoa-New Zealand's total population. The city stands at odds with the mythology of Aotearoa-New Zealand as a rural space.[12] From the outset of the British invasion, settlers and politicians saw the colony as a space in which to exercise a "better" settler colonialism, an almost utopian alternative to British life in other colonies or in Great Britain.[13] Edward Gibbon Wakefield (1796–1862) famously outlined this vision in a theory of systematic colonization that emphasized maintaining a perfect population density; the plentiful pastoral spaces of New Zealand would offer restoration and a better future to British settlers damaged by the flawed industrialized reality of Europe. These ideas crystallized in the new imperial imaginary for New Zealand, spurring settlement while also accelerating genocide, encouraging dispossession, and exacting drastic, often damaging, tolls on ecosystems.

Crucially, settler conceptions of pastoral, rural, and urban areas did not (and often still do not) align with Māori conceptualizations of land. The

[12] Bell, Claudia. 1997. "The 'Real' New Zealand: Rural Mythologies Perpetuated and Commodified." *The Social Science Journal* 34, no. 2: 145–58. 10.1016/S0362-3319(97)90047-1; Alam, Ashraful, and Etienne Nel. 2022. "Migration, Emerging Multiculturalism and Planning in Rural and Small Town Aotearoa New Zealand." *Australian Planner* 58, no.3–4: 132–42. 10.1080/07293682.2023.2169724.

[13] Gentry, Kynan. *History, Heritage, and Colonialism: Historical Consciousness, Britishness, and Cultural Identity in New Zealand, 1870–1940*. United Kingdom: Manchester University Press, 2015; Higgins, Katie W. 2017. "The Migrancy of Racial and Settler Imaginaries: British Migrants in Auckland, Aotearoa New Zealand." *Social & Cultural Geography* 20, no. 1: 87–108. 10.1080/14649365.2017.1347956.

deployment of these categories perpetuates settler-colonial exploitation, as evidenced by the continued enthusiasm for utilizing Aotearoa-New Zealand's pastoral spaces for settlement, agriculture, tourism, or film. The nostalgia for these spaces functions as a reminder that despite the failure of many of the spaces to restore former British subjects or create a model colony free of exploitation, nineteenth-century settler-colonial, pastoral ideals have by no means vanished.

However, in the mid-twentieth century, amid global economic pressures, shifting trade relationships, and a looming energy crisis, being perceived as rural became increasingly problematic for Aotearoa-New Zealand. One of the systems that the Public Works Department, which became the Ministry of Works Department (MWD) (1948), believed the electric grid would transform the rural spaces and develop industries in Aotearoa-New Zealand. Electric power authorities intended to do more than power the lights. Local politicians, power companies, and industry wanted the city and the surrounding boroughs of Auckland to grow and jumpstart existing industries so that more might be welcomed.

During the first half of the twentieth century, the North Island, due to settler-colonial land acquisition and geographic viability, maintained disjointed electric power supply networks primarily generated by hydropower.[14] By the 1960s, Aotearoa-New Zealand's Parliament and the Ministry of Works Department decided that the construction of a large thermal power station could create the possibility of greater energy independence and a more versatile electric power infrastructure. The result was the Huntly Power Station, Aotearoa-New Zealand's largest coal-fired power station to date. The architects of the Huntly scheme hoped the station (and perhaps others like it) would rejuvenate its mining community, retooling their rural space to service the expansion of an urban space. Ministry of Works Officials, regional politicians, environmental activists, mining families in Huntly, and the Waikato-Tainui all became engaged in suggesting how the existing energy landscape of Huntly could be reconfigured, ultimately uniting settler agendas and clarifying existing Waikato-Tainui distrust of and resistance to the government.

[14] Helen Reilly, *Connecting the Country: New Zealand's National Grid 1886–2007* (Aotearoa New Zealand: Steele Roberts, 2008), 70.

Huntly in Aotearoa-New Zealand's History

Huntly has a complex history that predates its emergence as a coal-mining center. Before European colonization, the Ngāti Mahuta, a hapū (clan) of the Tainui iwi, lived along the banks of the Waikato River.[15] Known as Rāhui Pōkeka, the settlement provided abundant resources, including fertile soils and access to waterways for trade and travel.[16] The arrival of Pākehā (non-Māori in New Zealand) in the nineteenth century led to significant disruptions, particularly during the New Zealand Wars (1845–1872), when British colonial forces sought to suppress Māori resistance to land confiscations.[17]

Governor George Grey (1812–1898) ordered an invasion of the Waikato during the Waikato Campaign of 1863–1864, displacing many Waikato-Tainui people, who were forced southward. As part of Raupatu (confiscation), vast tracts of Māori land, including those around Rāhui Pōkeka, were seized and redistributed to European settlers. During the late nineteenth and early twentieth centuries, Huntly transformed from a military outpost into a coal-mining town.[18] At the same time, agricultural and cattle-ranching industries expanded, fueling settler pastoral mythologies.

Although Māori likely knew of rich seams of coal, their use of coal (te waro) is unclear. Large-scale extraction began with European mining operations. For decades, coal affected the lives of all who lived in and around Huntly. The industry flourished in the late nineteenth century; coal works expanded across the country and became one of its primary energy sources used until the postwar years. Coal powered the trains that Prime Minister Julius Vogel hoped would connect the country and increase the potential flow of steamships into Aotearoa-New Zealand. Coal and gas-powered domestic industries like freezing, meat packing, lighting, and heating all relied heavily on coal.

[15] Vincent O'Malley, *The Great War for New Zealand: Waikato 1800-2000* (Wellington, NZ: Bridget Williams Books, 2016.).

[16] Corey Ruha, Sylvia Tapuke, and Aneika Young, *Mai te rangi kit te whenua, mai te whenua ki te rangi: A Kaupapa Māori literature review identifying land-based values and actions to benefit freshwater systems* (Rotorua, NZ: Scion, 2021): 21.

[17] Richard Boast and Richard S. Hill, *Raupatu: The Confiscation of Māori Land* (Melbourne, AU: Rotorua, NZ Victoria University Press, 2010.).

[18] Jane Legget, "Mining the Mining Museum on New Zealand's North Island: Rich Veins of Dissent," in *Mining Heritage and Tourism*, Michael Conlin and Lee Jolliffe, eds. (Routledge, 2010), 57–58.

By 1898, mining in Huntly produced around 100 tons of coal a day. It also brought challenges, including labor disputes and deadly accidents, most notably the 1914 Ralph Mine disaster, which claimed 43 lives. Other small industries, like brick making, flourished in Huntly, but it remained mainly a mining town well into the mid-twentieth century when opencast mining began on both the east and west sides of the Waikato River in 1945. As demand for coal declined, Huntly's economy diversified, yet its historical roots as a rural mining town remained.[19] Today, residents of Huntly continue to wrestle with their heritage and industrial past, reflecting the complex history of colonial expansion, resource extraction, and resilience in Aotearoa-New Zealand.[20]

Electricity and the Priority of the Urban in the Late Twentieth Century

Beginning in the middle of the twentieth century, the debate surrounding the construction of the Huntly Power Station saw Huntly's image transformed from that of just a coal-mining town into one that could play a meaningful role in "modernizing" the country's electricity generation infrastructure.[21] The growing population of the North Island, primarily in Auckland, convinced the government to focus on developments there, much to the dismay of boosters from the South Island. During the mid-twentieth century, Auckland was expanding to include larger suburbs and industries that required more electric power.[22] The Huntly Power Station project was positioned as a crucial enabler of Auckland's growth. Even in 1950, a decade prior to formal talks within the Ministry of Works Department (MWD) about constructing a power station in Huntly, the

[19] Legget, 59–60.

[20] Andrew McRae, "Ralph Mine Remembered 100 Years On," Radio New Zealand, September 12, 2014, Last Accessed September 9, 2024. https://www.rnz.co.nz/news/national/254388/ralph-mine-remembered-100-years-on Len Richardson, *Coal, Class, and Community: The United Mine Workers of New Zealand, 1880–1960* (Auckland, NZ: Auckland University Press, 2013), 126; recently, some residents of Huntly have lamented energy policies that diminished the use of coal. Richard Walker, "When Coal Was King, Huntly Boomed," *Waikato Times*, October 24, 2024. Last Accessed September 9, 2025. https://www.waikatotimes.co.nz/culture/360461682/when-coal-was-king-and-huntly-boomed/

[21] Reilly, *Connecting the Country*, 58–59.

[22] Paul Moon, *Auckland: The Twentieth Century Story* (Aotearoa Books, 2023), 204.

Minister of Public Works, William Goosman, saw Huntly as a solution to looming electric power shortages.

> The Minister's investigation into the Central Waikato Power Board's scheme to build a fuel station on the Huntly coalfields — a site ideally central to cater for the North Island demand and with no transport difficulties or costs for fuel—will be awaited with interest.[23]

The drive for more power only grew in the 1960s. For example, T. E. Kelly, president of the Electric Supply Authority Engineers' Institute, highlighted the geographical discrepancy between hydropower production and electricity demand:

> Of the possible water power available, the South Island would contribute 80 percent and the Auckland Province would use one third and the North Island three-fifths more.[24]

Although Huntly became the preferred site to meet the needs of growing power demand on the North Island and to lessen the demand on the hydropower production on the South Island, the electricity was for Auckland. This shift in perspective is notable—healthy rural towns, while desirable, were not the ultimate goal. Instead, the desirability of a power station in Huntly underscored how urban-centered the focus on New Zealand's future had become. Auckland, transformed from a sprawling conurbation into a compact and congested cityscape by the 1940s, embodied the ideal urban space.[25]

Since its founding in 1840, Auckland served as a military outpost, capital city (before being replaced by Wellington), major port, and eventually hub of the North Island's roads and railways, and manufacturing center, serving rural industries and thus helping to make possible the mythology of rural New Zealand. But as time went by, the importance of manufacturing and population growth shifted that balance, with the urban area becoming a larger focus for development. Auckland's emergence as a fast-growing urban center in the twentieth century depended on the growth

[23] "Electric Power-Ever Increasing Demand: Potential Sources Investigation by Minister," *Waihi Daily Mail*, March 3, 1950, 2.

[24] "Stupendous Power Needed," *Press*, September 23, 1965, 3.

[25] Miles Fairburn, "The Rural Myth and the New Urban Frontier: An Approach To New Zealand Social History, 1870–1940." *New Zealand Journal of History* 9, no. 1 (1975): 3–21.

of the country's electric power infrastructure.[26] Electric power began to appear in Aotearoa-New Zealand during the 1880s. Initially, small cities and industries wielded the electric light to advertise, encourage further settlement, and appeal to ideas of modernity that aligned with the British Empire. While many early systems, like the power station at the Phoenix Mine (1887) or the electric works at the Waihi Gold Mining Company, later the Horahora Power Station (1913), were funded by specific industries, eventually the government shifted the authority away from private enterprise.[27] Electro-Motive Power Act of 1896 made it impossible for anyone to use waterways for electric power generation without the government's permission.[28] Early twentieth-century geographical surveys such as the 1904 *Hay's Report on New Zealand Water-Powers* highlighted the value of hydropower arguing that "the utilisation of all the hydraulic power available in the colony seems to be quite probable in the not very distant future ... the development of hydraulic power will shift present and create new industrial centres in the Colony."[29]

Aotearoa-New Zealand saw an explosion of electric power development during the early twentieth century as the Public Works Department, later the MWD, undertook major hydroelectric projects on the North and South Islands and developed transmission lines between the major power stations. For Auckland, the construction of the Arapuni Power Station (1929/32), transmission lines to other North Island Stations, and supplementary diesel generators would, after decades of construction, meet the city's demand for electricity. By the mid-1940s, following the Depression, material shortages created by World War II, and a series of earthquakes that crippled several of the country's dams, public works officials began to

[26] R. C. J. Stone, *From Tamaki-Makau-Rau to Auckland* (University of Auckland Press, 2001), 267; J.G. Culy, E.G. Read, and B.D. Wright. "The Evolution of New Zealand's Electricity Supply Structure." Chapter. In International Comparisons of Electricity Regulation, edited by Richard J. Gilbert and Edward P. Kahn, 312–65. Cambridge: Cambridge University Press, 1996.

[27] Reilly, *Connecting the Country*, 34.

[28] F.T. M. Kissel, "The Organization of Electricity Supply in New Zealand," *Journal of the Institution of Electrical Engineers* 77, no. 463 (1935): 12; for a longer history of the relationship between electrification and settler colonialism in Aotearoa-New Zealand, see, Nathan Kapoor, *Power Colonialism: Electrification, Extraction, and Empire in Aotearoa New Zealand* (University of Pittsburgh Press, Forthcoming).

[29] Peter Seton Hayes, Superintending Engineer of the Public Works Department, "New Zealand Water-Powers," *Appendix to the Journal of the House of Representatives*, 1904, Session 1, D-01a.

worry whether the country's existing infrastructure, dams dependent on rain, could keep up with demand. Officials sought new power sources and control mechanisms to improve power output and diversify energy sources. The plans for a power station in Huntly intensified during this scramble.[30] And notably, the vitality of urban growth as an end in itself intensified along with it.

The wealth of coal at Huntly had been used for decades to justify an electric power station there, in keeping with the colonial government's framing of Aotearoa-New Zealand's resources as tools to support the rural and urban industries that made the colony valuable. In the early 1920s, as a result of the difficulties surrounding the construction of the dam at Arapuni, the Auckland Power Board explored the option of building a thermal station at Huntly but reasoned that the fluctuation of fuel prices presented too much risk. They argued that Auckland and the agricultural industries surrounding it, as one author put it in the *New Zealand Herald*, "the Empire's Dairy Farm," required the limitless power from the Waikato River.[31] However, natural disasters, unpredictable rainfall, and a growing demand for electricity proved that hydropower from the Waikato was not limitless; some officials preferred steam, citing the expansion of steam power in the United States and Britain as more diverse and secure systems for generating electric power.[32] The disruption of power in Auckland and the rocky operation of Arapuni featured frequently in the press, especially as the North Island production remained deficient when compared with that of the South Island.

By this time, Huntly's mines were seeing a major decrease in the demand for coal, which in turn led to declining population since it was coal that had motivated the establishment of the town after the Waikato Wars. What's more, Huntly, like many other small towns in Aotearoa-New Zealand, felt the economic blows of major worldwide price decreases in wool and dairy products. These threats worsened as Great Britain began its long entrance into the European Economic Community (1973), a move that saw Aotearoa-New Zealand's special trade privileges with Great Britain disappear and a subsequent decline in Aotearoa-New Zealand's import and export industry with Great Britain and Europe. All of this coupled with the looming Oil Shock of 1973 left Aotearoa-New Zealand

[30] Reilly, *Connecting the Country*, 141.

[31] "Supply of Power," *New Zealand Herald*, November 13, 1923, 35 (supp.).

[32] "Steam Generation of Electricity," *Dominion*, September 25, 1941, 10.

industries, cities, and politicians concerned about the country's self-sufficiency.[33] More electric power produced by domestic resources meant more electric power for transportation, communications, industry, and domestic consumption.

In addition to wrestling with how best to supply the country's growing electric power demands with local resources, between 1950 and 1965, Ministry of Works officials sought to connect the country with high-voltage lines. This reorganization drew further attention to the limits of hydropower on the North Island. In 1950, the Chief Electrical Engineer at State Hydro argued that by 1959, Aotearoa-New Zealand would reach its maximum hydroelectric capabilities and began advocating for high-voltage DC links across the Cook Strait to connect both islands. This debate highlights the government's worries about the North Island's deficiency in domestically produced power. The South Island's burgeoning industries and hydroelectric potential left officials searching for an answer to the North Island power problem; Huntly became one of the answers.[34]

Bridging Timelines with the Huntly Power Station

> Our plans to hasten exploration were well-based. We have valuable alternatives to oil, and we are redoubling our efforts in this direction … Obviously we cannot rely entirely on oil, if it is going to be used as a political weapon.[35]

As suggested in this quote from T. M. McGuigan, Minister of Energy, in an interview about the Energy Crisis of 1973, New Zealand government officials consistently tied domestic-powered electricity production to autonomy. Debates about infrastructural support for electrification between the 1940s and 1980s embed particular narratives that connect the past, present, and future, specifically the trajectory from colony to Dominion to self-sufficient nation. Ambitions for national growth were

[33] Moira Smith, "Culture Shock: The Legacy of the 1960s Power Generation Schemes in Aotearoa New Zealand." *Architectural History Aotearoa* 19 (2022):61–82.

[34] "S.I. Industry Urged," *Press*, October 9, 1969, 13.

[35] Notably, McGuigan positions the oil crisis as an outsider's issue rather than situating New Zealand within a global energy economy. This demonstrates how public works officials have garnered support for their initiatives by centering self-sufficiency as essential for the colony and later lone Dominion in the South Pacific. Cedric Mentiplay, "Comment from the Capital: Some Problems of Power–A Forecast and a Warning," *Press*, November 26, 1973, 14.

firmly and continuously predicated on the logic of dispossession that informed settler colonialism in Aotearoa-New Zealand from the first arrival of colonists. Between the confiscation of Māori land following the Aotearoa-New Zealand Wars and the construction of the Huntly Power Station, the Crown disregarded Māori claims and exploited Māori land and resources to generate electric power.

During the 1920s, for example, the government passed legislation to make Lake Taupo and the Waikato River Crown property, despite being recognized by Māori as the ancestral territory of the Ngāti Tūwharetoa iwi. Since the invasion of the British, many Māori sought to resist through the appointment of a sovereign. In 1858, Pōtatau Te Wherowhero led the Kīngitanga or Māori King Movement. The movement became a bastion of military and legislative resistance for many Māori, including the Waikato-Tainui, many of whom lived in and around Huntly. Settlers referred to this region as King Country because it was understood to be the last refuge of Māori resistance to the settler state. Indeed, Dame Te Atairangikaahu (1931–2006), the Māori Queen from 1966 to 2006, resided in the Waahi Pa (residence) in Huntly during her reign. By seating a major power station on the Waikato to provide electric power primarily for the settler industries of Auckland, the Huntly Power Station cannot be disentangled from decades of government land confiscation and disruption of Māori communities. In siting the Huntly plant as they did, public works officials implicitly prioritized settler-colonial objectives, suggesting that the politics of settlement that had informed major decisions from colonial times ought to still hold sway for the common good.

In 1969, the Energy Planning Committee of NZED submitted a report that contained the first concrete proposal for a power station at Huntly, although less definite plans had been drafted as early as the 1950s. The bid suggested building a 1000 MW power station, comparable to but slightly smaller than contemporaneous projects like the Grand Coulee Dam in the United States and the Volga Dam in Russia. It would deliver more electric power than all four dams on the Waikato combined. Although a nuclear plant was considered, ultimately, the authors and NZED agreed on a thermal power station that would employ more local power sources.[36]

[36] "Huntly 1000MW Thermal Power Station (1973) [Report]." Huntly Power Station. Date 1973-1975. (Archives Reference No: AAUM W4043 BOX 69, Record Number: ENV 8/2 Part: 2). (Wellington, New Zealand). Archives New Zealand.

The government considered a few locations for the station, but Huntly mayor, T. Gavin, and the City Council effectively promoted their town. As a collapsing coal-mining town of approximately 5000 in 1969, they sold NZED on the availability of coal, regional natural gas production, and the water from the Waikato River for necessary cooling. For many people, including Huntly officials, Waikato-Tainui and settlers alike, both the significant construction effort and the power station promised employment and growth which could regenerate the town.[37]

Huntly's hard times had started circa 1920, with declining demand for coal. Like so many other towns around Auckland focused on extractive industries, their dependence on one commodity left them economically vulnerable. Declining demand was in no small part due to the expansion of the electric power network. People and industries preferred electric power. Moreover, fewer people and mines were required due to changes in mining technology such as mechanized draglines, power shovels, and loading equipment. But the Minister of Electricity Leslie Walter Gandar (1919–1994) promised that should Huntly be chosen as the site for the new thermal station, "the mining of this coal would be the biggest operation undertaken in this country."[38]

The drive to build at Huntly was local, concerned with creating jobs and reviving the economy, but the project must also be understood as a response to the global pressures created by the oil crisis. In commemorating the official contract of the station's construction, the Minister of Electricity McGuigan defended hiring a Canadian engineering firm for the boiler design by reminding attendees that "all four boilers will be designed for firing with both coal and natural gas. The use of indigenous resources was part of the government's policy of greater self-reliance in energy." Self-reliance frequented the writings and policies of colonial politicians in Aotearoa-New Zealand from the colony's founding. As a "model" colony, Aotearoa-New Zealand was intended to be productive and restorative for the British settler population at minimal cost to the empire. These attitudes remained cornerstones of national identity even as Aotearoa-New

[37] Direct quotations from Waikato-Tainui during the construction of Huntly are rare. Tom Winston Fookes, a lecturer in Geography at the University of Waikato, did extensive research including interviews, but his published works lack transcripts. T. W. Fookes, *Huntly Power Project: A Description* (University of Waikato, 1981).

[38] "Massive Mining for Huntly Power," *Press*, May 5, 1972, 2.

Zealand transitioned from a colony into a nation. The explicit evocation of self-reliance in McGuigan's remarks deeply resonated with many people.

Despite the narrative of development that saw the power station as a source of rescue for Huntly and a self-sufficient Aotearoa-New Zealand, other narratives portrayed the plans for a power station in a less favorable light. These narratives suggested that instead of the continuity of restoration, the nation's new infrastructure projects required a sharp break with the past. Growing environmental activism, reflecting both global and local concerns, informed one such narrative. One project whose harms ignited public concern was the Manapouri Power Scheme, in which the government planned to raise a lake by tens of feet and merge it with another lake to provide power to the Tiwai Point Aluminum Smelter.[39]

The public feared what destruction might materialize at the proposed station at Huntly and how much power the government ought to have to pursue development unchecked.[40] Public works officials now recognized that new electrification projects would have to meet new environmental regulations. In addition to the authority of local power boards and NZED, the Huntly power station now also had to satisfy the new Minister of the Environment. Materially, this meant the power station's completion required an environmental impact statement (EIS), the first government infrastructure project to require an EIS in New Zealand.

Some MWD officials treated the EIS as a hindrance while opponents of the station believed the statement was not done seriously or expansively enough to actually change construction plans. This conflict highlights the national government's focus on producing more electric power and its disregard for countervailing concerns for the environment; the object was not to provide power *to* Huntly but to draw it *out of* Huntly for the productive industries of places like Auckland on the North Island. The consequences to Huntly were of secondary importance.

A second narrative emerged challenging whether the Huntly station as imagined should really be seen as a "common good" of urban Auckland or Aotearoa-New Zealand generally. The self-reliance so promoted by McGuigan and others was always built on the assumption that Māori

[39] The Manapouri Scheme is often cited as the project responsible for jumpstarting the environmental movement in New Zealand. Brierley, Gary, Ian Fuller, Gary Williams, Dan Hikuroa, and Alice Tilley. 2022. "Re-Imagining Wild Rivers in Aotearoa New Zealand" *Land* 11, no. 8: 1272. 10.3390/land11081272

[40] Thomas O'Brien, "Fragmentation or Evolution? Understanding Change Within the New Zealand Environmental Movement." *Journal of Civil Society* 9 (2013): 287–99.

needs, rights, and authority were dispensable or easily sacrificed. Despite local excitement about the power station and its potential employment, some Māori voices called for a different option. For example, Robert Mahuta (1939–2001), a prominent Māori politician and brother of the Māori Queen, drew significant attention to the harms posed by a coal-fired power station on the Waikato.[41] He pointed out that the construction unfairly impacted Māori and blatantly disregarded their concerns. He cited the carelessness of governing bodies who approved the station despite worries about how the temperature of the Waikato River could increase, the social tensions and insults hurled at Māori in Huntly who "slowed progress," the disruption of the Rakaumanga School (a Māori school), and the suppression of environmental reporting that should have shut down the project.

Activists, academics, and community members accused NZED and the government of duplicitous activities, including withholding data about negative environmental impacts, abusing earlier town and planning legislation to circumvent local political authority, and refusing to protect Māori community spaces because they were not considered "public" spaces. In an analysis of the project, Robert Mahuta, working as the Director of the Centre for Māori Studies and Research at Waikato University, summarized letters from several public works officials and raised the following concerns:

> When questioned as to the potential noise generated from the station site, Mr. Jones replied somewhat facetiously, "It will be like wind rustling through the trees." Yet from May 1974 it is envisaged that there will be at least 8 months of solid pile driving, several years of pop-riveting, and once completed, 20 to 30 years of station operation! The most recent statement issued by Mr. Jones … is aimed at subduing criticism expressed by the local people and implies that the rights and wished of the local Māori community

[41] Sir Robert Te Kotahi Mahuta remains one of the best historical sources for understanding the Waikato-Tainui and their relationship with the Crown, due to his family connections with the Māori royal family and education. He completed a master's in anthropology, produced dozens of papers, and tirelessly advocated for the Waikato-Tainui. He helped reclaim some half a million hectares from the Crown and $170 million from a Treaty of Waitangi Settlement. However, many Waikato-Tainui hold mixed feelings about his work because he also encouraged wild investment schemes later in his career, which some argue lost the Waikato-Tainui millions. Angela Gregory, "Sir Robert Mahuta: A Flawed Colossus," *New Zealand Herald*, February 1, 2001, Last Accessed September 23, 2025, https://www.nzherald.co.nz/nz/sir-robert-mahuta-a-flawed-colossus/OQGY572VFECADLUAYHKGXXOZ5I/

are totally insignificant when the conflict with the national demand for electricity.[42]

The list of accusations is lengthy and includes such problems as coal ash collection, the negative impact of construction, and the immediate environmental impact of such a massive power station.[43] It deserves more analysis than we can provide here. However, the debates surrounding the Rakaumanga School are worth highlighting because they reveal how the Huntly project expanded electrification by expanding Waikato-Tainui disenfranchisement.

In 1896, the Rakaumanga Primary School formed under the Native Schools Act (1867). Over the coming decades, despite poor support from the government and suppression of Te Reo Māori and Māori culture, this school became a strong community-building space for the Waikato-Tainui in Huntly.[44] Even after the act was dissolved in 1969, the Rakaumanga School continued to function as a space for many Waikato-Tainui children and families. The school sat directly in the path of the power station and the roads needed for construction; NZED and the Ministry of Works Department determined to move it. The Waikato-Tainui backlash became a centerpiece of their protests of the station. Even though NZED and the Education Board claimed the school was due to close and a new school would be built in West Huntly, the removal of the school for the power station made plain the government's disregard for Māori culture and the costs borne by the Māori of the government's vision of energy security. Every decision made by NZED and the government revealed a willingness to sacrifice Māori needs, rights, and connection to place. The school, its location, and the people who valued it did not contribute to the worth of Huntly for project boosters. Even political leaders were more concerned about losing the Māori vote than on the harms themselves, as the correspondence of the Labour Prime Minister Norman Kirk demonstrates:

[42] Robert Te Kotahi Mahuta, "Planning Issues and the Huntly Power Station," Centre for Māori Studies and Research, (1978), 11.

[43] For a longer description of this episode in the story of Huntly's construction, see Nathan Kapoor, *Power Colonialism: Electrification, Extraction, and Empire in Aotearoa New Zealand* (Pittsburgh University Press, Forthcoming 2026).

[44] Other schools served similar roles. Moon, Paul. 2019. "'To Teach the Natives English': James Pope's 1880 Native Schools Code." *History of Education* 48, no.5: 591–606. 10.1080/0046760X.2019.1585582.

> The Maori people who had such confidence in a Labour government are becoming disillusioned at the erosion of their rights and the compulsory buying of their land by the Government. Their environment is also being disturbed, and this is causing disruption.[45]

If advocates saw the Huntly Power station bridging the past and the present through its ongoing development of national self-sufficiency, opponents saw the continuation of dispossession. In 1978, following events like those at the school, Robert Mahuta asked the government:

> What kinds of society are we really trying to create in this country? Another America, another Japan? Here in Huntly the questions as far as the Maori people are concerned are – Progress, at what price? Have you a better world to offer than what we possess at the moment? We do not know. For over 100 years you have been making promises, today we are asking – prove it to us.[46]

Mahuta's objections demonstrate that not everyone in and around Huntly saw more electric power as the primary issue and certainly not as rescue. He and others saw the power station, and the reach of national public works officials, as little more than a continuation of settler colonialism. In critiquing the government's plans, opponents like Mahuta proposed a quite different way of bridging those timelines, one in which promises to the Māori carried weight and Māori spaces had value beyond the energy they could generate for Aotearoa-New Zealand's urban centers. For example, Waikato-Tainui critique pointed out that the station did not have to sit directly within these marae (meeting grounds) and that NZED could pursue a working relationship with them to advocate for their community.[47] They wanted a break with the past tactics of

[45] Norman Kirk to J. A. Walding, March 13, 1973, Huntly Power Station. Date 1973-1975. (Archives Reference No: AAUM W4043 BOX 69, Record Number: ENV 8/2 Part: 2). (Wellington, New Zealand). Archives New Zealand.

[46] Robert Te Kotahi Mahuta, "Planning Issues and the Huntly Power Station," Centre for Māori Studies and Research, (1978), 25.

[47] The meanings of marae shifted in Māori communities over time. The space can refer to the physical structure or complex built to house community functions, or a place of refuge and community extending around the structure and tied to the physical space the marae occupies. For the purposes of a history of the Waikato-Tainui in Huntly, the marae does refer to physical buildings, but it should also be understood for that place's long-running histori-

dispossession, opening a new avenue for respect and negotiation. With few exceptions, they did not get that break.

In 1977, NZED's response to such criticism was to award the Huntly Borough Council a $1 million grant to expand and improve public amenities, a tepid offer that made no concession to uprooted communities. They constructed a new public library and sports complexes to accommodate population growth and attract and keep people. Government projections promised that Huntly's population would reach 9–11,000 by mid-1977. In reality, the population only rose to 6200. NZED justified its electrification scheme as both a new supply of electricity and a source of development for Huntly. But development was still predicated on the acceptance and continuation of settler-colonial tactics of acquisition.

During the multiyear construction, despite consistent calls to reconsider the implications of the station, Ministry of Works officials, such as Robert Norman, continued to insist on its value. They argued that "shortages of power by 1978/79" would cripple the economy and "that deferment could place an obstacle in the way of the natural gas negotiation and almost certainly result in the government having to pay a high price for the gas."[48] For public works engineers, Huntly remained the preferred site because the station could burn local coal and, eventually, gas piped from the Maui Gas Field 307 km away in Taranaki. So the project barreled on with locals excited about what the power station might bring but also wary of what it might cost financially and environmentally in the long run. NZED continued its improvement efforts, including building a $1 million esplanade and shopping front along the river in 1978. Not all were convinced by the results. One Hamilton journalist in the *Press* said, "Huntly will get a facelift, but travelers passing through may not notice."[49] Development initiatives were always half-hearted. Improving rural spaces was always secondary to the urban need for power.

Such efforts did not distract NZED's most serious critics. Concerned parties, such as the Waikato-Tainui Māori in Huntly, often led by Robert

cal association with active resistance to the Crown. Paul Tapsell, "Marea and Tribal Identity in Urban Aotearoa/New Zealand," *Pacific Studies* 25, no. 1 (2002); 6.

[48] Robert Norman, "Proposed Huntly Power Station Assessment Environmental Impact," Officials Committee for the Environment, June 14, 1973, 7. Archives New Zealand (Wellington) R128876. Environmental Impact Reports: Huntly Power Station; Robert Norman, *You Can't Win 'Em All: Confessions of a Public Works Engineer* (Wellington: Slide Rule Press, 1997).

[49] "Facelift for Huntly," *Press*, December 19, 1978, 7.

Mahuta, claimed that environmental impact statements ignored the socio-economic consequences of the station while also underplaying the environmental factors. One attempt by the MWD to understand and mitigate the power station's impact was the Huntly Social and Economic Impact Monitoring Project, contracted with Waikato University and led by geographer Tom Winston (T.W.) Fookes. He framed the issue in terms of "fairness."[50] Was it fair to ask the small town of Huntly to suffer short and long term? Would the power station bring the benefits promised? If not, what should be done? Indeed, for the Waikato-Tainui Māori, the Huntly Power Station's reliance on land confiscation policies was nothing more than the government's latest attempt to continue traumatizing the indigenous peoples of Waikato.

Conclusion

Stories of rural rescue through infrastructure frequently hide a more complex reality. Although some disputes about Huntly highlighted strong environmental concerns, significant opposition from the Māori community focused more on pushing back against long-standing practices of dispossession and disregard for Māori perspectives on the significance of marae and other spaces than on opposition to the power plant per se. Robert Mahuta, for example, supported a power station but insisted that other sites in the region were more suitable than Huntly.[51] For Mahuta and others, using rural spaces to provide for the needs of Auckland was not in itself a problem. Retooling energy infrastructure to create greater self-sufficiency might even have been a shared value between some opponents and NZED. Instead, the continuity of settler-colonial tactics that would disrupt the center of Waikato-Tainui life and history was the problem. More than simple NIMBYism, their objections spoke to their desire to use electric power infrastructure to bridge timelines by breaking with the harmful politics of the past. Doing so required that Māori conceptions of space and its meaning be taken seriously. They refused to accede to Pākehā interpretations of place and the relational imaginaries that determined how places fit into a larger framework of the "common good." The Māori did not carry their point; NZED connected the past to the present

[50] T. W. Fookes, *Huntly Power Project: A Description* (Waikato University, Monitoring Social and Economic Impact, 1981), 1–3.

[51] Robert Mahuta to P.W. Blakely, March 9, 1973, 3.

in their preferred way, recapitulating settler-colonial priorities by embarking on construction of the station in 1974.

Ultimately, the Huntly Power Station only modestly rescued Huntly from decline. The population grew by just a few thousand people, and efforts at urbanization fell well short of their boosters' ambitions. It succeeded in demonstrating how settler tactics of dispossession continued to operate in modern times through the construction of infrastructure. Perhaps the best evidence for the utility of the Huntly Power Station in this regard was its status as poster child for the National Party's "Think Big" initiative. Robert Muldoon (1921–1992), the Prime Minister between 1975 and 1984, adopted "Think Big" as the title for economic policies that directed the government to borrow heavily from foreign governments to invest in "big" schemes to boost exports, reduce reliance on imports, especially oil, and raise wages through tax cuts.[52] Politicians like Muldoon and Ministry of Works officials modeled the land acquisition and borrowing used in Huntly to organize schemes like the Motuni Synthetic Oil Plant (1987), the Clyde Dam (1992), and the Waitara Methanol Plant (1983). By the mid-1990s, historians, economists, politicians, and the public generally agreed that Think Big initiatives were a fiasco, creating inflation, wage stagnation, and high national debt.

Huntly, employing about 500 people, produced electric power to feed Auckland and another growing city, Hamilton. As new hydroelectric and geothermal power plants were developed around New Zealand, Huntly, now part of Genesis Energy, became a power source.[53] At its peak in 2010, the Huntly Power Station met about 17% of Aotearoa-New Zealand's electric power demand.[54] In 2018, in response to Labour Prime Minister Jacinda Ardern's low-carbon emissions initiatives, Genesis Energy began discussing phasing out Huntly or replacing existing fuel sources with biofuels.[55] While solar and battery initiatives remain underway, climate shifts

[52] "More for Workers, Development for Business," *Press*, November 26, 1980, 1

[53] Parsons-Bronckerhoff, "2011 NZ Generation Data Update," Ministry of Economic Development, January 26, 2012, 11.

[54] This data was produced by Genesis Energy.

[55] Grant Bradley, "Coal No Longer King at Huntly Power Station," *New Zealand Herald*, August 5, 2015, Last Accessed September 23, 2025, https://www.nzherald.co.nz/business/companies/agribusiness/coal-no-longer-king-at-huntly-power-station/2VI6Y4ZXPXPQ5I5PPYIQ62PSA4/

and global instability in energy markets have reignited talks of keeping Huntly open until 2035.[56]

Many Waikato-Tainui see the Huntly Power Station as an institution responsible for employing many family members and, quite literally, connecting them to the rest of New Zealand. For the Waikato-Tainui, resisting the construction of a power station was part of a larger project. Following the establishment of the Waitangi Tribunal, a commission that makes recommendations to the government based on alleged breaches to the Treaty of Waitangi, the Waikato-Tainui fought for acknowledgment of the illegal confiscation of their land and resources. The Resource Management Act (1991) created a new arsenal to protect Aotearoa's resources. In 1995, Mahuta reached a settlement with the Crown, granting the Waikato-Tainui their claims over the Waikato River. In 2008, these negotiations finally resulted in the Kīngitanga Accords, which granted the Waikato-Tainui an apology, a legislative vehicle through which they and the Crown could negotiate, and Te Mana o Te Awa (Spiritual Authority, Protective Power, and Prestige of the River).[57] As the station relies on the Waikato River for cooling, they must cooperate with the Waikato-Tainui to address threats to the river. The process is uneven, but the Waikato-Tainui continue to challenge the power structures that informed the construction of the power station, seeking a future that disrupts settler-colonial dispossession.

Bibliography

Archives

Archives New Zealand n.d.
Alexander Turnbull National Library n.d.

[56] Jamie Gray, "Huntly Power Unit May Stay Operational until 2035, Genesis Confirms," *New Zealand* Herald, April 23, 2025, Last Accessed, https://www.nzherald.co.nz/business/companies/energy/huntly-power-unit-may-stay-operational-until-2035-genesis-confirms/LGRKYZQYKVEZPG4QWRP7XV4HSY/

[57] Waikato-Tainui Raupatu Claims (Waikato River) Settlement Act (2010); *The Kīngitanga Accord*, August 22, 2008, 3–4.

Books, Articles, Chapters

Alam, Ashraful, and Etienne Nel. "Migration, Emerging Multiculturalism and Planning in Rural and Small Town Aotearoa New Zealand." *Australian Planner* 58 (3–4) (2002): 132–42. doi:https://doi.org/10.1080/07293682.2023.2169724.

Bell, Claudia. 1997. "The 'Real' New Zealand: Rural Mythologies Perpetuated and Commodified." *The Social Science Journal* 34, no. 2: 145–58. doi:https://doi.org/10.1016/S0362-3319(97)90047-1;

Boast, Richard and Richard S. Hill, Raupatu: *The Confiscation of Māori Land.* Melbourne, AU: Victoria University Press, 2010.

Brierley, Gary, Ian Fuller, Gary Williams, Dan Hikuroa, and Alice Tilley. 2022. "Re-Imagining Wild Rivers in Aotearoa New Zealand" *Land* 11, no. 8: 1272. https://doi.org/10.3390/land11081272

Cater, Casey P. Regenerating Dixie: Electric Energy and the Modern South. Pittsburgh, PA: University of Pittsburgh Press, 2019.

Culy, J.G., E.G. Read, and B.D. Wright. "The Evolution of New Zealand's Electricity Supply Structure. Edited by Richard J. Gilbert and Edward P. Kahn. *International Comparisons of Electricity Regulation.* Cambridge: Cambridge University Press, 1996: 312–65.

Fairburn, Miles. "The Rural Myth and the New Urban Frontier: An Approach to New Zealand Social History, 1870–1940." *New Zealand Journal of History* 9, no. 1 (1975): 3-21.

Fookes, T. W. *Huntly Power Project: A Description.* Waikato University, Monitoring Social and Economic Impact, 1981.

Gentry, Kynan. *History, Heritage, and Colonialism: Historical Consciousness, Britishness, and Cultural Identity in New Zealand, 1870–1940.* Manchester: Manchester University Press, 2015.

Higgins, Katie W. "The Migrancy of Racial and Settler Imaginaries: British Migrants in Auckland, Aotearoa New Zealand." *Social & Cultural Geography* 20 no. 1 (2017): 87–108. doi:https://doi.org/10.1080/14649365.2017.1347956.

Howe, Cymene, Jessica Lockram, Hannah Appel, Edward Hackett, Dominic Boyer, Randal Hall, Matthew Schneider-Mayerson et al. "Paradoxical Infrastructures: Ruins, Retrofit, and Risk." *Science, Technology, & Human Values* 41, no. 3 (2016): 547-565.

Jasanoff, Sheila and Sang-hyun Kim. *Dreamscapes of Modernity: Sociotechnical Imaginaries and Fabrications of Power.* Chicago: University of Chicago Press, 2015.

Kale, Sunila. "Structures of Power: Electrification in Colonial India." *Comparative Studies of South Asia, Africa and the Middle East* 34, no 3 (2014): 454-475.

Kapoor, Nathan. *Powering Colonialism: Electrification, Extraction, and Empire in Aotearoa New Zealand* Pittsburgh, PA: University of Pittsburgh Press, 2026

Kissel, F.T. M. "The Organization of Electricity Supply in New Zealand." *Journal of the Institution of Electrical Engineers* 77, no. 463 (1935): 12.

Kline, Ronald. *Consumers in the Country: Technology and Social Change in Rural America*. Baltimore: Johns Hopkins University Press, 2002.

Larkin, Brian. "The Politics and Poetics of Infrastructure", *Annual Review of Anthropology* 42 (2013): 327-343.

Legget, Jane. "Mining the Mining Museum on New Zealand's North Island: Rich Veins of Dissent," in *Mining Heritage and Tourism*. Edited by Michael Conlin and Lee Jolliffe. London: Routledge, 2010: 57-58.

Mahuta, Robert Te Kotahi. "Planning Issues and the Huntly Power Station," Centre for Māori Studies and Research, 1978.

McRae, Andrew. "Ralph Mine Rembered 100 Years On," Radio New Zealand, September 12, 2014, Last Accessed September 9, 2024. https://www.rnz.co.nz/news/national/254388/ralph-mine-remembered-100-years-on

Moon, Paul. *Auckland: The Twentieth Century Story*. Auckland, NZ: Aotearoa Books, 2023: 204.

Moon, Paul. 2019. "'To Teach the Natives English': James Pope's 1880 Native Schools Code." *History of Education* 48, no. 5: 591–606. doi:https://doi.org/10.1080/0046760X.2019.1585582.

Muru-Lanning, Marama. *Tupuna Awa: People and Politics of the Waikato River*. Auckland: Auckland University Press, 2016.

Muru-Lanning, Marama. "The Analogous Boundaries of Ngaati Mahuta, Waikato-Tainui, and Kiinitanga." *The Journal of the Polynesian Society* 120, no. 1: 9–41.

Norman, Robert. *You Can't Win 'Em All: Confessions of a Public Works Engineer*. Wellington: Slide Rule Press, 1997.

O'Brien, Thomas. "Fragmentation or Evolution? Understanding Change Within the New Zealand Environmental Movement." *Journal of Civil Society* 9 (2013): 287–99.

O'Malley, Vincent. *The Great War for New Zealand: Waikato 1800–2000*. Wellington, NZ: Bridget Williams Books, 2016.

Parsons-Bronckerhoff. 2012. 2011 NZ Generation Data Update. Ministry of Economic Development, January 26, 2012, 11.

Pisani, Donald. "Federal Reclamation in the American West in the Twentieth Century." *Agricultural History* 77, no. 3: 391-419.

Reilly, Helen. *Connecting the Country: New Zealand's National Grid 1886–2007*. Aotearoa New Zealand: Steele Roberts, 2008: 70.

Richardson, Len. *Coal, Class, and Community: The United Mine Workers of New Zealand, 1880–1960*. Auckland: Auckland University Press, 2013: 126.

Ruha, Corey, Sylvia Tapuke, and Aneika Young, *Mai te rangi kit te whenua, mai te whenua ki te rangi: A Kaupapa Māori literature review identifying land-based values and actions to benefit freshwater systems.* Rotorua, NZ: Scion, 2021: 21.

Seton Heyes, Peter. n.d. Superintending Engineer of the Public Works Department, 'New Zealand Water-Powers'. *Appendix to the Journal of the House of Representatives* 1904, Session 1, D-01a.

Smith, Moira. "Culture Shock: The Legacy of the 1960s Power Generation Schemes in Aotearoa New Zealand." *Architectural History Aotearoa* 19 (2022): 61-82.

Stone, R. C. J. *From Tamaki-Makau-Rau to Auckland.* Auckland: University of Auckland Press, 2001: 267.

Tapsell, Paul. "Marea and Tribal Identity in Urban Aotearoa/New Zealand," *Pacific Studies* 25, no. 1 (2002).

Waikato-Tainui Raupatu Claims (Waikato River) Settlement Act (2010) 2008. The Kīngitanga Accord, August 22, 2008.

Whittle, Jo. "Into the Backyard: Huntly Power Station and the History of Environmentalism in New Zealand." *Environment and Nature in New Zealand* 8 (1) 2013: 7-38.

CHAPTER 7

Manufacturing Plus: Sociotechnical Imaginaries, Interpretative Flexibility, and the Ascent of Taiwan's Semiconductor Foundry Model

Ling-Fei Lin

Introduction

To borrow Herbert A. Simon's famous thought experiment regarding the mythical visitor from Mars:[1] What would a Martian see if it observed Earth from space through a telescope capable of recognizing our industrial categories—manufacturing, services, and so on? The Martian might be astonished to learn that making chips is classified as manufacturing, while making food in a restaurant is not. Is it because chips are physical objects? But isn't food also something tangible, something made? Why, then, is one called manufacturing and the other service? It might further observe

[1] Herbert A. Simon, "Organizations and Markets," *Journal of Economic Perspectives* 5, no. 2 (Spring 1991): 27–29.

L.-F. Lin (✉)
Indiana University Bloomington, Bloomington, IN, USA
e-mail: linglin@iu.edu

J. L. Lieberman, A. Mohsin (eds.), *Technological Humanities*,
https://doi.org/10.1007/978-3-032-24741-4_7

that these terms are antiquated and no longer capture what is happening in Earth's industrial worlds for a long time.

Oftentimes, the boundary between low- and high-value sectors is permeable and socially constructed. The notion that manufacturing is a "low-value" activity and services are "high-value" has been widespread, but this dichotomy has been challenged through multiple ways to understand what manufacturing could mean. For example, Morris Chang (張忠謀), the founder of today's semiconductor giant, Taiwan Semiconductor Manufacturing Company, Ltd. (TSMC; 台灣積體電路製造股份有限公司; 台積; 台積電), stressed that his business is a dedicated foundry in order to elevate the status or value of his business. In 1998, he even stated, "Chipmaking is no longer a manufacturing business—it's a service business". In 1995, he explained that what TSMC provided was infrastructure, not merely contract manufacturing (代工).[2] In other words, while many stakeholders in the high-tech economy had flattened the idea of manufacturing, Chang emphasized its multidimensional nature.

Chang's perspective is an influential one: he was called "the Father of Wafer Foundry" by Fabless Semiconductor Association and was elected by *Time* magazine and CNN as one of "the Top 26 Most Influential CEOs" in the world in 2001. His company, TSMC, accounted for more than 34% of global chip production, and over 70% of the pure-play foundry (純晶圓代工) market in 2024.

Chang first sought to overturn the low-value stereotype on manufacturing with better terminology. He used terms such as *dedicated foundry* (晶圓專工) and *pure-play foundry* (純晶圓代工) to emphasize the entanglement of service and manufacturing. These names have become much more widely used in the press and media since the mid-1990s, although the Chinese terminology could still lead the public to treat it as a form of contract manufacturing linked to a lower-value image. Here, to avoid reproducing the hierarchy that imagines service as inherently superior to manufacturing, I call their business model *manufacturing plus*. This term shows that this model still focuses on materially manufacturing or making chips, while remaining open to other service possibilities.

In this chapter, I argue that the role of manufacturing is open to multiple imaginaries and interpretative flexibility, a term I discuss at greater

[2] Morris Chang (張忠謀), *Autobiography of Morris C. M. Chang, Vol. 2: 1964–2018* (Chinese: 《張忠謀自傳下冊:1964–2018》) (Taipei: CommonWealth Publishing Group, 2024).

length below. In Taiwan's case, it took about a decade after its initial partnership with RCA (Radio Corporation of America) for this production openness to become relatively stabilized as the pure-play foundry model—a form that has transcended conventional understandings and practices of production before the 1970s.

This essay primarily focuses on the years 1974 to 1997. The year 1974 is significant as it marked the Taiwanese government's first initiative measures concerning the semiconductor industry. TSMC was founded in 1987 during this period, and the year 1997 marked TSMC's tenth anniversary—an occasion on which founder and CEO Morris Chang described Taiwan's integrated circuit (IC) industry as standing at a "crossroads." He used the crossroads metaphor to urge Taiwan to redirect its IC industry even though it had gained some success.[3] By 1996, Taiwan already ranked fourth worldwide in semiconductor output value, largely due to its two major pure-play chipmakers, TSMC and United Microelectronics Corporation (UMC). While a new wave of sociotechnical imaginaries and geopolitical actions emerged in the early twenty-first century, this essay focuses on the initial period when the pure-play foundry model was invented and stabilized.

The Start

To understand how Taiwan reached a *manufacturing plus* status, we must first understand the historical context in which high-tech manufacturing became a central part of the country's economy. One well-known narrative suggests that Taiwan's place in this global high-tech arena began with a morning meeting on February 7, 1974, Hua Fei (費驊), when the General Secretary of Taiwan's Executive Yuan, met with Sun Yun-Suan (孫運璿), Taiwan's Minister of Economic Affairs, Wen-Yuan Pan (潘文淵), Chairman of TAC (the Technical Advisory Committee) and four other government high officials in a traditional Taiwanese breakfast store. The aim of this meeting was to decide if Taiwan could develop semiconductor industry for Taiwan's economic transformation. They finally decided to develop integrated circuit (IC) production and transfer foreign

[3] Ling-Fei Lin (林玲妃), "Taiwan's IC Industry, Standing at the Crossroads" (Chinese: 《處於十字路口的台灣IC工業》), *China Times*, September 26, 1996. This transcript of Morris Chang's talk was also published in TSMC's internal magazine.

technology for it.[4] However, this famed breakfast was not just a dramatic one-time event, but the culmination of experiments and plans that would take shape over time and require professional maneuvering.

Some narratives about Taiwan's emergence as a semiconductor powerhouse emphasize the technology transfer from RCA, a US-based electronics company.[5] These technology-centered explanations are partly valid, only addressing hardware and technology. Such discussions overlook the social factors that fostered the growth and vitality of Taiwan's semiconductor industry. To extend these incomplete histories, I draw on two approaches from Science and Technology Studies: sociotechnical imaginaries and the social construction of technology (SCOT). Together, they help explain TSMC's rise in Taiwan across two phases.

Inspired by earlier studies on imagination,[6] Sheila Jasanoff and Sang-Hyun Kim introduced in 2009 the concept of sociotechnical imaginaries to contrast two national visions of nuclear power.[7] They show that the United States cast the atom as "atoms for peace" and as a potentially runaway technology requiring strict "containment," whereas South Korea imagined nuclear technology as "atoms for national development," a symbol of national progress and technological capability.[8] That is, the same

[4] See Chin-Tay Shih (史欽泰), Chen Tien-Chih (陳添枝), and Wu Shu-Min (吳淑敏), From the Periphery to the Core: How Taiwan's Semiconductor Industry Became the Heart of the World (Chinese: 從邊緣到核心:台灣半導體如何成為世界的心臟》), Kindle ed. (2025); Ju-Shin Chang (張如心) and Pan Wen Yuan Foundation (潘文淵文教基金會), *Silicon Story of Taiwan,* (Chinese: 《矽說台灣》) (Taipei: CommonWealth Publishing Group, 2006), 83–85; Yi-Yan Hong (洪懿妍), *Innovation Engine: ITRI, The Helper of the Success of Taiwan Industry* (Chinese: 《創新引擎:工研院:臺灣產業成功的推手》) (Taipei: CommonWealth Magazine, 2003), 32–34; John A. Mathews and Dong-Sung Cho, *Tiger Technology: The Creation of a Semiconductor Industry in East Asia* (Cambridge University Press, 2000), Chapter 4.

[5] The emphasis on the RCA project is typically provided by participants and by ITRI itself, while some figures, such as Morris Chang, seldom or never stress this history with RCA.

[6] Charles Taylor's *Modern Social Imaginaries* (2004), which focuses on a shared sense of legitimacy about social life, is an inspiration to their sociotechnical imaginaries. George E. Marcus edited a book, *Technoscientific Imaginaries*, in 1995, which collects texts from 24 scientists in the late twentieth century about how they imagined science or technology that they developed or participated in, and its associated morals and ethics.

[7] Sheila Jasanoff and Sang-Hyun Kim, "Containing the Atom: Sociotechnical Imaginaries and Nuclear Power in the United States and South Korea," *Minerva* 47, no. 2 (2009): 119–146.

[8] Jasanoff and Kim, "Containing the Atom," 121, 131.

technology was promoted with different imaginaries or conceptions of the sociotechnical life.

Jasanoff and Kim define national sociotechnical imaginaries as "collectively imagined forms of social life and social order, reflected in the design and fulfillment of nation-specific scientific and/or technological projects. Imaginaries, in this sense, both describe attainable futures and prescribe futures that states believe ought to be attained," and consider that "the capacity to imagine futures is a crucial constitutive element in social and political life."[9] Jasanoff later extended the concept of sociotechnical imaginaries to: "collectively held, institutionally stabilized, and publicly performed visions of desirable futures, animated by shared understandings of forms of social life and social order attainable through, and supportive of, advances in science and technology."[10] The concept highlights the entanglement between social agency on one hand and technological futures on the other, envisioning how these technologies would be closely integrated with social order and collective life. I consider sociotechnical imaginaries particularly useful for analyzing the early, broad vision articulated by technocratic officials and young elites; SCOT (discussed later) becomes more effective once that vision stabilizes and is translated into specific, innovative manufacturing models that facilitate industrial transformation. Together, the two frameworks provide a conceptual apparatus for understanding how Taiwanese actors imagined—and actively wove—their own interpretations of manufacturing into their ideals and into their sense of Taiwan's social and global situatedness.

State Planners

Taiwan's government and technocratic high officials had imagined the importance of industrial manufacturing to Taiwan's economy well before the breakfast meeting. Taiwan occupied a strategic position to "contain" communist China during the Cold War—particularly after the start of the Korean War (1950–1953). During this time, Taiwan received American aid of around 1.4 billion US dollars, averaging about 5% to 10% of Taiwan's GDP annually.[11] Later, the United States found out that

[9] Jasanoff and Kim, "Containing the Atom," 120, 122.

[10] Sheila Jasanoff, "Future Imperfect: Science, Technology, and the Imaginations of Modernity," in *Dreamscapes of Modernity: Sociotechnical Imaginaries and the Fabrication of Power*, ed. Sheila Jasanoff and Sang-Hyun Kim (University of Chicago Press, 2015), 4.

[11] Jinqing Liu (劉進慶), *An Economic Analysis of Taiwan After the War* (Chinese: 《台灣戰後經濟分析》) (Taipei: Ren-Jian Publishing Company, 1992), 350–351. See also Maurice

Taiwan spent too much of the American aid on national defense. In other words, despite that large influx of funds, the Taiwanese economy did not grow during these years.[12] Under pressure from the United States, and because of its own small domestic market, Taiwan's government began to incentivize export-oriented industries, even establishing an export processing zone in 1966, which was claimed to be a pioneering project in the world.[13] Before the early 1960s, Taiwan had virtually no electronics industry.[14] Consequently, imagining and defining Taiwan as an "export machine" (as I call it) for the world became crucial to its shift toward the electronics and semiconductor industries, given the island's small domestic market. This export-oriented imaginary distinguished Taiwan from economies such as Japan and China, whose large home markets fostered very different industrial mindsets.

At this point in 1974, when officials gathered at the breakfast meeting, Taiwan faced a series of diplomatic frustrations and other challenges. In the early years after the separation of Mainland China and Taiwan, there seemed to be hope that the Nationalists would regain Mainland China's regime one day. But year after year, countries shifted diplomatic recognition to Mainland China. In 1971, Taiwan was expelled from the United Nations; in 1972, Japan broke off the diplomatic relations with Taiwan; the first global oil crisis occurred due to the Arab-Israeli War. All these frustrations and setbacks drove the Taiwanese government to adopt a strategy of self-reliant economic development.[15]

Wen-Yuan Pan, then a research lab director at RCA in the US, also the chair of TAC, proposed that it was time to upgrade Taiwan's electronics industry from labor-intensive to technology-intensive, and integrated circuits seemed to be the most crucial and most influential technology among many. But the key question was: *How can we develop the IC industry in Taiwan?* Pan suggested that, since time was very crucial, it was better to purchase technology from America. After detailed discussions, they agreed to develop the IC industry with substantial government support and US technology.

Scott, "Taiwan's Trade Development" (Chinese: 「臺灣的貿易發展」), trans. Chao-Wu Lin, in *The Paper Selections of Foreign Trade Development in Taiwan (Continuation)* (Chinese: 《臺灣對外貿易發展論文集(續編)》), ed. Chi Schive (Taipei: Linking Books, 1994), 6–35.

[12] Jinqing Liu, *An Economic Analysis of Taiwan After the War.*

[13] *The Export Processing Zone of the Republic of China: Reviewing the Past and Mapping Out the Future* (Taipei: Ministry of Economic Affairs, 1996), 4.

[14] Chang, Wu, and Yu, in Chang and Yu, *Vitality*, 38.

[15] Hong, *Innovation Engine*, 27.

During the Cold War, the arms race between the United States and the Soviet Union was intense. In the 1950s, spending for defense research and development in the United States even surpassed its wartime peak.[16] Many high-tech projects were largely supported by the military.[17] The US military played a key role in developing computers, atomic energy, and telecommunications technologies. In contrast, the relationship between military support and the early development of semiconductors is much in dispute. Nonetheless, the early development of semiconductors was still associated with federal sponsorship and military and space projects' procurements, even though the military's role was only supplementary.[18]

In general, the United States outshines others in the contribution of basic technologies in the history of semiconductors. However, the US promotes a sociotechnical imaginary of the high tech as separate from manufacturing, justifying and encouraging the offshoring of labor-intensive production work to other countries. This opened a niche—and created space—for East Asian countries to enter high tech. Within East Asia, Taiwan developed a distinctive pathway. As Mathews and Cho argue, Taiwan patiently developed a "build-up of private sector capabilities using the leverage of the public sector."[19]

Taiwan was indeed characterized by strong state intervention at the time. Nonetheless, without the collaboration of the private sector, industrial experts, and young graduates, this history might not have unfolded at all. I also argue that in the mid-twentieth century, a prevailing global

[16] Stuart W. Leslie, *The Cold War and American Science* (Columbia University Press, 1993), 1.

[17] See Leslie, *The Cold War and American Science*. For the computer's early development, see Paul N. Edwards, *The Closed World: Computers and the Politics of Discourse in Cold War America* (MIT Press, 1996), Chapters 1–3.

[18] For example, see Daniel I. Okimoto, Takuo Sugano, and Franklin B. Weinstein, eds., *Competitive Edge: The Semiconductor Industry in the U.S. and Japan* (Stanford University Press, 1984), 79, Chapters 2 and 4; for a detailed description and literature of this dispute, see Christophe Lécuyer, *Making Silicon Valley: Innovation and the Growth of High Tech, 1930–1970* (MIT Press, 2005), Chapter 4. On the notion that most basic technology was invented in private labs, see Okimoto, Sugano, and Weinstein, *Competitive Edge*, 81–82. On the notion that the US Department of Defense played a larger role through procurement policies, see Lécuyer, *Making Silicon Valley*, Chapter 4; Miller also indicates how the early IC industry depended on the order of the military. See Chris Miller, *Chip War: The Fight for the World's Most Critical Technology* (Scribner, 2022). On Japan's role in development, see Lécuyer, *Making Silicon Valley*, 14–34.

[19] Mathews and Cho, *Tiger Technology*, 313.

sociotechnical imaginary in Taiwan cast the island as an "export machine"—specifically, a machine for manufacturing goods for the world. This idea emerged against the backdrop of advanced countries' visions of a post-industrial society, as well as Taiwan's own efforts to develop its economy and "high tech" amid a tense political climate overshadowed by China.[20]

Beginning in 1966, the Chinese Institute of Engineers (CIE) in the US (美洲中國工程師學會) convened the Modern Engineering and Technology Seminars (METS, 近代工程技術討論會), representing a bottom-up mechanism from overseas experts.[21] Later, Taiwan's government began systematically seeking advice from overseas on its technological development blueprint and strategic direction. In 1974, the US-based Technical Advisory Committee (TAC, 科技顧問會議) was founded by the government to focus on technology development in Taiwan. In 1980, to partially institutionalize and elevate the TAC, the Executive Yuan established a Science and Technology Advisory Group/Meeting (STAG, 行政院科技顧問小組) to be directly under the premier, replacing the TAC's advisory function in a former and higher-level way.[22] The METS and the TAC were both comprised of overseas Chinese technology experts, and the STAG had an even higher level of foreign members, including US industrial executives such as Patrick E. Haggerty, the retired chairman of Texas Instruments; Bob O. "Boe" Evans, one of the principal leaders behind IBM's landmark System/360 and a key figure to advocate IBM's compatible computers; and Ian Munro Ross, a transistor pioneer who also served as President of Bell Labs for 12 years. These meetings and advisory mechanisms shaped how Taiwan envisioned its role and progress within a broader international context.

[20] Ministry of Economic Affairs, Republic of China (Taiwan), *The Export Processing Zone of the Republic of China: Reviewing the Past and Mapping Out the Future* (Taipei: Ministry of Economic Affairs, 1996), 4. Taiwan's government claimed that Taiwan's EPZ was the only such zone in the world at the time, since before Taiwan's EPZ, there were only free trade zones and industrial zones globally. EPZs differed from free trade zones and industrial zones, combining the strengths of both earlier models.

[21] This CIE was established in 1912 in China, but after China and Taiwan's split in 1949, the organization was also split, with headquarters in Taipei, Taiwan, and overseas branches in the US, Japan, and so forth.).Sec https://www.cie.org.tw/p/412-1000-91.php?Lang=zh-tw (Accessed 30 June 2026)

[22] See Shih et al., *From the Periphery to the Core: How Taiwan's Semiconductor Industry Became the Heart of the World* (Chinese: 從邊緣到核心:台灣半導體如何成為世界的心臟》), Kindle ed. (2025); Chang and Pan Wen Yuan Foundation, *Silicon Story of Taiwan*.

Meanwhile, during this period, Taiwan's government was typically very efficient at translating blueprints and plans into execution. As Ding-Yuan Yang avers, "Back in that time Minister Sun had slowly institutionalized the process of the whole technology of development."[23] Yang, one of the original 19 talents sent to RCA for technology transfer in chipmaking, noted that these meetings tended to state the conclusion first, followed by further discussion, after which a formal plan and conference record would be produced. The semi-governmental research and development organization Industrial Technology Research Institute (ITRI) would then "propose and sign a project contract with the Ministry of Economic Affairs, and proceed with its execution."[24] That is, these foreign or overseas experts would periodically gather and provide options to help outline a future blueprint for Taiwan's technological development, which was then implemented by ITRI. The transfer of RCA's technology (1976), the establishment of the demonstration factory in Taiwan (1977), the creation of UMC (1980) based on that factory, and the later founding of TSMC (1986) were among the key outcomes of these mechanisms.

When deciding to develop Taiwan's IC industry, state technocrats first relied on foreign expertise. Taiwanese government officials could not envision the technical elements of semiconductor production, but they knew Taiwan had to transform its economy, and they believed that the high-tech market would buoy their goals. Despite the intentionality of their vision, there was never one single vision of the sociotechnical imaginary surrounding the development of Taiwan's IC industry. Initially, legislators and the National Science Council (NSC) opposed the establishment of ITRI and the semiconductor development plan. Legislators argued that the creation of the ITRI would blur the line between public and private interests: it would use government funding but would not be subject to normal government oversight, and once this precedent was set, it would create many worries about similar cases in the future.[25] The NSC also objected to the development of semiconductors in Taiwan, probably due to the exclusion of other projects for the sake of the RCA project.

[23] D. Y. Yang, *Taiwanese IT Pioneer: D. Y. (Ding-Yuan) Yang*, 12. https://archive.computerhistory.org/resources/access/text/2012/02/102745994-05-01-acc.pdf

[24] Yang, *Taiwanese IT Pioneer*, 13.

[25] Kuo-li Chang (張國立), *Biography of Sun Yun-suan* (孫運璿傳) (Taipei: Commonwealth Publishing Group, YEAR), chap. 11.

In 1975, Jerry Chu, an NSC consultant and former Vice President of Honeywell, wrote a detailed report at the invitation of Taiwan's NSC, in which he strongly questioned the feasibility of transferring semiconductor technology from abroad. Within the four main categories of IC applications, he suggested that Taiwan should "backward integrate" (from the computer) to develop Taiwan's electronics industry; he imagined that forward integration of the semiconductor itself would have an extremely high chance of failure, supposing that Taiwan could not compete with the US counterparts.[26] Then-Chairman of Taiwan's NSC, Li-an Chen (陳履安), a rising political star and son of former Vice President Chen Cheng (陳誠), also opposed the founding of TSMC, proposing instead that the US partner (Arizona State Science Park) build a clean room module in Taiwan's science park so that Taiwan would not have to invest directly in semiconductors.[27] This dispute about how best to approach the intensified. According to Chang, K. T. Li (李國鼎, a minister without portfolio overseeing science and technology) even said he would resign from the government if this issue could not be solved.[28] K. T. Li has long been credited as one of the key government officials behind the founding of TSMC.

ITRI played an important role in shaping the development of Taiwan's IC industry. It incubated UMC and TSMC—the two largest IC foundry companies in Taiwan. UMC spun off from ITRI's RCA project in 1980, and seven years later, TSMC also spun off from ITRI's VLSI (Very Large Scale Integration) project. ITRI also provided both technologies and human resources for the IC industry in Taiwan.[29]

Jasanoff and Kim argued that imaginaries encapsulate "an organized field of social practices," which serve as a "key ingredient in making social order."[30] In precisely this way, the Taiwanese government actively reimagined their niche in the global order. Under the leadership of Yu, Sun, and K. T. Li—and after a few decades of effort by the government, political leaders, industrial leaders, and skilled professionals—Taiwan transformed from a poor agrarian region into an export-driven manufacturing society.

[26] Chang, *Autobiography*, and Shih et al., *From the Periphery to the Core.*

[27] Chang, *Autobiography*, 341.

[28] Chang, *Autobiography.*

[29] Bin-Win Peng, Houn-Gee Chen, and Bou-Wen Lin, "A Taiwan Research Institute as a Technology Business Incubator: ITRI and Its Spin-Offs," *Comparative Technology Transfer and Society* 4, no. 1 (2006): 5.

[30] Quoted in Jasanoff and Kim, "Containing the Atom," 122.

Indeed, the intentionality behind this particular sociotechnical imaginary might explain why the 1974 breakfast meeting is so renowned in Taiwanese culture. Although this story began with technology transfer, the Taiwanese officials in this story were innovators. Although they did not develop a new kind of chip, they created an environment that fostered Taiwan's distinctive IC model.

Technological progress was widely regarded as essential for national development, yet Taiwanese government officials with strong technical backgrounds did not succumb to technological determinism—waiting for technology to change their circumstances. Instead, they carefully crafted an approach that integrated advice from overseas experts with their own understanding of their unique socioeconomic position and homegrown engineering talent. Together, these elements paved the way for Taiwan's semiconductor industry to achieve global success—a high status which began as an imaginary long before it became a reality.

The Young Talents

State planners' sociotechnical imaginaries were not the only perspectives that mattered; young technical elites also helped fuel the fire to start the IC engine in Taiwan. These young talents carried government officials' imaginaries forward, re-envisioning desirable futures for Taiwan's society through hands-on engagement in high tech. Very importantly, they were strongly motivated by the national mission.[31] In the US, where they observed the huge difference between the living standard in the US and that in Taiwan during that time. They returned to Taiwan with the technical expertise and motivation to initiate change – and to transform imaginaries into practices.

Though the major endeavors on semiconductors with governmental support did not start until 1974, some industrial seeds had already taken root by that time. The first concerns skilled professionals. As early as 1964, Chun-Yen Chang (張俊彥), Jui-Fu Chang (張瑞夫), and Shuang-Fa Kuo

[31] Shih recalled that they were so motivated when Sun Yun-Suan came to talk with them before they departed for the US, as they were young and meeting with such a high official was unthinkable. Sun mentioned how significant their job would be, and they must succeed for Taiwan. Chin-Tay Shih (史欽泰), *Taiwanese IT Pioneer: Chintay Shih*, 9. In the CHM file, they use Chintay as his first name, but Shih uses Chin-Tay himself. Both names refer to him. To avoid confusion with another interviewee, Stan Shih, Acer's main founder, I spelled out their full names for clarity.

(郭雙發) set up the first Taiwan semiconductor lab in National Chiao-Tung University (NCTU).[32] In 1966, they produced Taiwan's first integrated circuit. Later, many of their graduates joined electronics companies with foreign investment in Taiwan or started up their own companies, creating a talent pool for Taiwan.[33]

After sending the request for proposals (RFP) to 28 American companies, the government selected RCA among seven companies that submitted proposals.[34] With the agreement struck, they sent a training cohort of 36 engineers (later often called the "seed troops") for the RCA project. The first team consisted of 19 men, all in their late twenties to early thirties. These elite engineers had either studied at Taiwan's top universities or possessed rare expertise and specific experience in high-tech sectors. Three of the four regional leaders held newly earned doctoral degrees in electrical engineering from Princeton University.[35] Ding-Yuan Yang (楊丁元) served as head of the entire team and leader of the New Jersey region, which focused on design. Chintay Shih (史欽泰) led the Ohio region, concentrating on the production process. The third regional team, based in California and led by Ching-Chu Chang (CC Chang, 章青駒), learned testing technology, while the fourth team in Florida, led by Gene Sheu (許健), who had earned his Ph.D. in solid-state physics from the University of Iowa, conducted research on equipment. [36] Clearly, the ambition of this group of high-level young engineers was not confined to IC production. Their interests also encompassed testing, equipment (closely tied to production), and IC design.

Shih said that he, CC Chang, and Yang all graduated from Princeton. Because they were more familiar with life in the US, the three of them had an important responsibility beyond their technical-training goals: to chauffeur their peers during the early months of the RCA transfer project until

[32] National Science and Technology Council (NSTC). 2018. Taiwan Semiconductor Industry Pioneer - Chang Chun-yen Passes Away, Ministry of Science and Technology to Request Presidential Commendation ("臺灣半導體產業先驅-張俊彥辭世 科技部將呈請總統褒揚"), press release, October 16, 2018.

[33] Chang and Pan Wen Yuan Foundation, *Silicon Story of Taiwan*, 72-78.

[34] Shih et al. *From the Periphery to the Core*; Shih, *Taiwanese IT Pioneers*, Shih mentions that in total, Taiwan issued a request for proposal to 28 US companies, but many just rejected them, as some of them were too small to do such a project.

[35] https://cc-winbond.azurewebsites.net/, click the pdf titled "用對產業的熱情 寫下半導體業發展史," and https://50th.itri.org.tw/history/semiconductors/1/ (Accessed 1 Mar 2026).

[36] Ibid.

everyone obtained their driver's licenses. They also cooked for themselves because they were unaccustomed to American food.[37] Their time living together in the US fostered strong camaraderie, which served them well as they became core leaders of Taiwan's semiconductor industry in the decades to come. Shih commented that RCA factories did not really need people with their expertise and talent, yet the Taiwan team had a different purpose: to study the whole production process. RCA soon came to treat them as equal partners, and eventually as fellow problem-solvers and not just trainees. While in the US, the Taiwan team supplemented their RCA training by making connections with other companies. For instance, they learned quality and reliability from Hewlett Packard and mask technology from International Material Research (IMR).[38]

This RCA project involved comprehensive learning. Shih said, "Taiwan's government was not dedicated to technology transfer at all, and technology transfer was virtually done [just] at the enterprise-level. At the enterprise level, the technology transfer was about the provision of the so-called turnkey solution, which would be provided as a fixed solution that means no alternation were allowed... Technology transfer should allow new element additions, through which new solutions can be developed to solve specific problems and can be further advanced."[39] In other words, although they learned from RCA, they were not passive partners learning just about a static process.

Indeed, the Taiwanese team helped raise yield rates—something they could not have accomplished without a deeper understanding of semiconductor manufacturing. Shih mentions that yield rates can determine if IC production succeeds or not, as high yield rates will lead to low cost; however, low yield rates can be caused by many different factors. The Taiwanese team could solve this problem only because these young talents worked hard to understand the technology.[40] Ding-Hua Hu (胡定華), the Taiwan-based director who managed the connection between the RCA project and ITRI's demo factory, also attributed high yield rates to their talent team.[41] In the socioeconomic context of the oil crisis, the Taiwanese team did not just have an exclusively technical goal. They wanted to contribute to the sociotechnical futures beyond what Taiwanese government officials

[37] Shih, *Taiwanese IT Pioneers*, 10.
[38] Hu, *Taiwanese IT Pioneers*, 16.
[39] Shih, *Taiwanese IT Pioneers*, 11.
[40] Shih, *Taiwanese IT Pioneers*, 8.
[41] Hu, *Taiwanese IT Pioneers*, 17.

were plannning for.[42] They also wanted Taiwan's semiconductor industry to become self-dependent.

Interviews with these team members often translate the RCA project into their hopes for Taiwan's economy and society. Their sociotechnical imaginary linked high tech to high living standards, signaled by Taiwan's self-reliance and its growing economy. This demonstrates that although the stated goal of the RCA project was mainly to learn the semiconductor production technology, the participants shared much broader ambitions.[43] The team departed for the US in 1976, helped set up Taiwan's IC demonstration factory in 1977, and then produced the first batch of ICs for electronic watches in 1978. This move propelled Taiwan to become the world's third-largest exporter of electronic watches and laid a solid foundation for Taiwan's IC industry.[44] This talented team did not disappoint the government and Taiwan's society. Most team members later made substantial contributions to Taiwan's semiconductor industry. Many rose to leadership positions across Taiwan's IC industry: Robert Tsao (曹興誠) was the founding and long-term chairman of UMC, Fan-Cheng Tseng (曾繁城) served as TSMC's vice chairman, while Ming-kai Tsai (蔡明介) continues to chair MediaTek, now also a fabless IC giant. Others included Parkson Chen (陳碧灣), who headed Taiwan Mask Corporation, Shih, who directed ITRI, and DY Yang and CC Chang, who led Winbond Electronics. Several others founded their own companies or made significant contributions.

The RCA project team leader, Young, gave credit to the Taiwanese government during that time.[45] He said that Japan's development of DRAM (dynamic random-access memory) was a national strategy, and so was Taiwan's and China's. To him, these countries believe that "if they hope to develop a self-sustainable autonomous industry, they have to control the technologies of semiconductors."[46] According to him, "the semiconductor industry is one of the great pillars in the technological civilization in human history." He even "Thank[s] the Heavens for having blessed Taiwan' that seized the chance to develop the technology of com-

[42] Hu, *Taiwanese IT Pioneers,* 17.
[43] Shih et al., *From the Periphery to the Core.*
[44] https://50th.itri.org.tw/history/semiconductors/1/.
[45] Yang, *Taiwanese IT Pioneer.*
[46] Yang, *Taiwanese IT Pioneer*, 27.

plementary metal-oxide-semiconductors (CMOS) technology from such an early stage."[47] When transistor-transistor logic (TTL) was the dominant technology, Taiwan bet on CMOS, and TSMC grew with CMOS as it became mainstream in the industry.[48]

Additionally, several things are notable in this history. First, in the technology transfer project, what Taiwan acquired was not the most advanced technology, and the enhancement of yield rates was not due to the technology itself, because Taiwan's yield rates eventually surpassed RCA's. (Indeed, RCA later proposed a merger that Taiwan rejected.) Second, Taiwan chose to cooperate with RCA because it was the only company willing to transfer design knowledge as well. Taiwan insisted on acquiring design capability because it wanted to avoid then-standard products such as TTL (Transistor-Transistor Logic) and focus instead on ASICs (Application-Specific Integrated Circuits), which required design capability. [49] Later, ITRI even added the idea of also incorporating the capability for project management.[50]

One unforeseen result was that employees trained through Taiwan's semiconductor sector have also contributed to the rise of China's IC industry. Although China had built its own semiconductor production systems since the 1950s, it was relatively lagging and considered a newcomer in the global semiconductor market in 2000. Yet, thereafter, it gained great momentum through aggressive strategies, including hiring former executives and employees from Taiwan to build its own pure-play company SMIC,[51] and pushed Taiwan's UMC and TSMC to build fabs locally for China's market notwithstanding Taiwanese governmental restrictions.[52] Overall, China gained both commercial markets and defense ICs' capability in a time when the West had looser restrictions on them. Thousands of senior employees and some executives, legally or illegally, went to work for

[47] Yang, *Taiwanese IT Pioneer*, 7.

[48] Shih, *Taiwanese IT Pioneers*; Shih et al., *From the Periphery to the Core*..

[49] See Shih, *Taiwanese IT Pioneer*, and Shih et al., *From the Periphery to the Core.*

[50] United Daily newspaper, page 2, 12/29/1975.

[51] Ming-Chin Monique Chu, "China's Defence Semiconductor Industrial Base in an Age of Globalisation: Cross-Strait Dynamics and Regional Security Implications," *Journal of Strategic Studies* 47, no. 5 (2024): 643–668, https://doi.org/10.1080/01402390.2023.2164852.

[52] Chu, "China's Defence Semiconductor."

China's semiconductor firms.[53] Clearly, machines cannot operate seamlessly by themselves. Engineers and managers with experience and expertise are equally indispensable.

Business Innovators: Reimagining and Reinterpreting Manufacturing

Semiconductor production cannot be understood without considering global sociotechnical imaginaries of manufacturing. Western sociotechnical imaginaries invested in the free market and imagined manufacturing as already relatively unimportant in their advanced society—and even as low-end activities after the 1970s. During that time, the high-value sector referred to fields such as service, information, and knowledge work—domains that aligned with the concept of the "postindustrial society," or the "third wave" and the "network society," all of which regarded manufacturing as a peripheral or subordinate sector.[54] Thus, many American and European companies saw little issue in outsourcing manufacturing to other countries.

This widespread Western sociotechnical imaginary of manufacturing played a role in configuring the global division of labor landscape and contributed to the emergence of TSMC and its innovative pure-play foundry business model in the silicon industry. However, the popular imaginary alone is not sufficient to explain the complex interrelationships that specifically shaped Taiwan's *manufacturing plus* model. Interpretative flexibility, a concept proposed by Trevor Pinch and Wiebe Bijker within their discussion of the social construction of technology (SCOT) approach, offers an important methodology at this juncture. Interpretative flexibility holds that technological artifacts or systems are developed, shaped, and used in multidimensional ways by different social groups which invest different meanings into the technology. Also, it further suggests that how different relevant social groups negotiate around this flexibility and eventually achieve closure, stabilizing the technology into a particular form.[55]

[53] Chu, "China's Defence Semiconductor."

[54] Daniel Bell, *The Coming of Post-Industrial Society: A Venture in Social Forecasting* (New York: Basic Books, 1973). Alvin Toffler, *The Third Wave* (New York: William Morrow, 1980); Manuel Castells, *The Rise of the Network Society*, 2nd ed., vol. 1 of The Inforation Age: Economy, Society, and Culture (Chichester, UK: Wiley-Blackwell, 2010). Originally published in 1996.

[55] Trevor Pinch and Wiebe Bijker, "The Social Construction of Facts and Artefacts: Or, How the Sociology of Science and the Sociology of Technology Might Benefit Each Other,"

The classical SCOT case studies focus more on competing or purpose-specific artifacts, or technological objects such as bicycles, automobiles, and bakelites, and problemitize them in a more microscopic way.[56] In the case of Taiwan's semiconductor industry, I extend the framework from an artifact or a technological configuration to an industrial ecology and an innovative business model.

In the early stages of Taiwan's IC industry, many social groups were involved, but professional technocrats, industrial entrepreneurs, and young technical talents were especially crucial. They seemed oriented toward a common broader direction, yet beneath the surface, they disagreed about how best to achieve the sociotechnical imaginary of Taiwan as a major producer of integrated circuits. In other words, interpretative nuances across relevant social groups further elaborate Taiwan's story in the early future-visioning–showing how different actors endowed manufacturing with different meanings in the first place.

In the case of Taiwan's semiconductor industry, some of the interpretative flexibility was rhetorical. A dominant interpretation of manufacturing as "low-value" led to the conditions where this work was generally outsourced. By considering manufacturing as unskilled labor, advanced economies—especially the United States—lost the ability to make "real goods" in integrated-circuit, personal-computer, and other electronics sectors: their domestic supply chains either shifted offshore or vanished altogether, eventually reconsolidating in Asia. This created an opportunity for Taiwan, although the dimensions of that opportunity were unclear at the time. In an oral history interview from 2007, Morris Chang described the situation in the following way:

> What strengths have we got? [T]he conclusion was very little. We had no strength in research and development, or very little anyway. We had no strength in circuit design, IC product design. We had little strength in sales and marketing, and we had almost no strength in intellectual property. The only possible strength that Taiwan had, and even that was a potential one,

in *The Social Construction of Technological Systems*, ed. Wiebe E. Bijker, Thomas P. Hughes, and Trevor Pinch (MIT Press, 1987).

[56] Ronald R. Kline, *Consumers in the Country: Technology and Social Change in Rural America* (Johns Hopkins University Press, 2000); Ronald R. Kline and Trevor Pinch, "Users as Agents of Technological Change: The Social Construction of the Automobile in the Rural United States," *Technology and Culture* 37 (1996); Wiebe E. Bijker, Thomas P. Hughes, and Trevor Pinch, eds. *The Social Construction of Technological Systems* (MIT Press, 1987).

> not an obvious one, was semiconductor manufacturing, wafer manufacturing. And so what kind of company would you create to fit that strength and avoid all the other weaknesses? The answer was pure-play foundry. So maybe you could call it the least evil choice.[57]

By choosing the lesser of two evils, what Morris Chang meant was that he did not have an inherently good option. In order to make the best of a bad situation, he responded to objections from Taiwan's NSC by piecing together an amalgam of approaches—a dynamic process that at times seemed forced into a specific direction. While serving as President of ITRI (1985–1987), Premier Kuo-hwa Yu (俞國華) asked Chang to "transfer research results into economic gains for Taiwan industry,"[58] aiming to leverage high-tech exports as Taiwan's ticket to the global stage. Looking back, Morris Chang appears to have used an elimination strategy to find Taiwan's particular focus in developing semiconductors in the very beginning, as the above statement shows. But this "elimination" strategy in the mid-1980s was not simply a matter of luck; it drew on his 30 years of experience in the semiconductor industry before that time.

Morris Chang's unique decision contrasted greatly with popular practices at that time. There were few fabless IC companies (or "design houses"), as the main model was an IC company which designed, manufactured, and tested the chips they needed in the house (called IDM, integrated design manufacturing). However, in the new pure-play or dedicated foundry model, players concentrated on making chips for others. Their clients include pure IC design houses or "fabless" companies, as well as IDMs that outsource part of their product manufacturing. *Foundry* is a term coined by one of Intel's founders, Golden Moore, to refer to "no design, only manufacturing," although foundry was just a secondary job in Intel, an IDM which could make chips for others when it had spare capacity. When Morris Chang invited Intel to invest in the new company TSMC, Intel declined, noting that they did not think foundries could be profitable. Undeterred, Morris insisted and called his new model as "dedicated foundry."[59]

[57] Morris Chang, Oral History Interview, https://www.semi.org/en/Oral-History-Interview-Morris-Chang.

[58] Morris Chang, *Oral History of Morris Chang*, interview by Allen Patterson (Computer History Museum, 2007), 12.

[59] Chang, *Autobiography*, 370-371. In fact, Dutch firm Philips was the only international partner willing to invest in TSMC, taking 27.6 percent, while Taiwan's government held 48.3 percent. Under repeated government pressure, several domestic companies reluctantly took the remaining shares, each holding only 1 to 5 percent; see Chang, *Autobiography*, chapter 17.

Therefore, the wafer foundry itself was not a totally new model.[60] However, no company in the world ran a "pure" foundry company (or called "dedicated silicon foundry") until TSMC did so in 1987. The interpretative flexibility Morris Chang capitalized on was more about an economic model than the technology itself. The shift to focus solely on the foundry was an innovation in this sector, and it enabled TSMC to provide stable, reliable, and customer-centered foundry service.[61]

After only a few years, this model proved to be both workable and profitable.[62] In fact, the TSMC model impacted the industry as a whole by creating a new division of labor: since there were pure professional manufacturing services now, those who could not afford the huge capital for a wafer fab could just start up a fabless IC design house instead, with much lower costs. Not long after TSMC's 1987 innovation, the pure-play foundry model fostered a worldwide boom in fabless IC design houses, and the whole IC ecology was largely changed.[63] Several became so successful that they rivaled traditional large integrated device manufacturers (IDMs, such as Intel and IBM) in both market share and revenue. The prime example is Nvidia—now the world's most valuable chip company—which grew in tandem with TSMC. Nvidia surpassed Microsoft's market capitalization in 2025, and achieved extraordinary gains despite, and arguably partly because of, its fabless strategy.

In his autobiography, Morris Chang also noted that one factor behind TSMC's move to the pure-foundry model was that three Silicon-Valley-based IC design firms founded by overseas Chinese—Mosel, Vitelic, and Quasel—approached Taiwan's government for funding to build their own wafer fabs on the island. Because the required investment was enormous, it also contributed to Chang's idea of building a common fab to serve all of them.[64] At the time, this idea was entirely novel, and it showed how hard it was for actors to step outside the prevailing imaginary that every chip company needed its own factory (Figs. 7.1 and 7.2). Although one motivation for creating the foundry model came from the needs of IC

[60] Mathews and Cho, *Tiger Technology*.

[61] Wu, "Stride forward to the world's number one dedicated wafer foundry business," in Chang and Yu (edit) Chap 5; Chang and Pan Wen Yuan Foundation, *Silicon Story of Taiwan*, 185–228; Peng, Chen, and Lin. Edited in Chang and Yu.

[62] Chang and Pan Wen Yuan Foundation, *Silicon Story of Taiwan*, 185–228.

[63] Between 1991–2000, the number of global fabless IC companies grew from 207 to 750, with total revenues increasing seven times. See Chang, *Autobiography*.

[64] Chang, *Autobiography*.

design customers, these were merely small startups. TSMC did not have stable or large firms to serve initially. Customers existed only potentially, making this one of the greatest risks for TSMC's massive investment—a risk that required bold vision to undertake.

By the mid-1990s, Chang had formed a much clearer vision, which included an ideal way to distinguish TSMC's uniqueness in the dedicated foundry business and to reshape what manufacturing could mean in the global IC industry. He indicated TSMC's success in part is on its awareness that chipmaking "is no longer a manufacturing business - it's a service business." [65] A small country like Taiwan can rely on being an "irreplaceable node in manufacturing" to occupy a position in the global supply chain.[66] In a 1995 talk, he already asserted that Taiwan could succeed by becoming the fab for the world, claiming that once Taiwan became an infrastructure provider in the global semiconductor industry, no one could ignore it.[67]

While Taiwan was being labeled as an assembly country, Morris Chang distinguished between labor-intensive assembly and knowledge-intensive manufacturing. He argued that a wafer foundry needs to combine cutting-edge technology and substantial capital investment; it is never low-value contract manufacturing. He also believed that Taiwan had to retain the (production) process know-how to sustain long-term leadership, and "Advanced manufacturing together with R&D is Taiwan's moat; design or branding alone cannot keep us ahead."[68]

He also complains in his autobiography and on other occassions that the Chinese translation of foundry was rendered directly as "晶圓代工" (wafer contract manufacturing). To him, foundry in English is neutral, but the Chinese word "代工" carries low-tech connotations. Although he added "professional" (專業) in front of it to become 專業晶圓代工, it still had a low-value connotation with the Chinese term 代工; he could not change the term that had been used conventionally in society.[69] Chang could not single-handedly shift the stereotyped language used in society to describe his work properly, but I argue that he made considerable efforts to create interpretative flexibility by rhetorically reframing the foundry as high-tech, high-capital, and high-service. Indeed, outsiders tend to flatten

[65] Chang, *Autobiography*.
[66] Chang, *Autobiography*.
[67] Chang, *Autobiography*.
[68] Chang, *Oral History of Morris Chang*.
[69] Chang, *Autobiography*, 370–371.

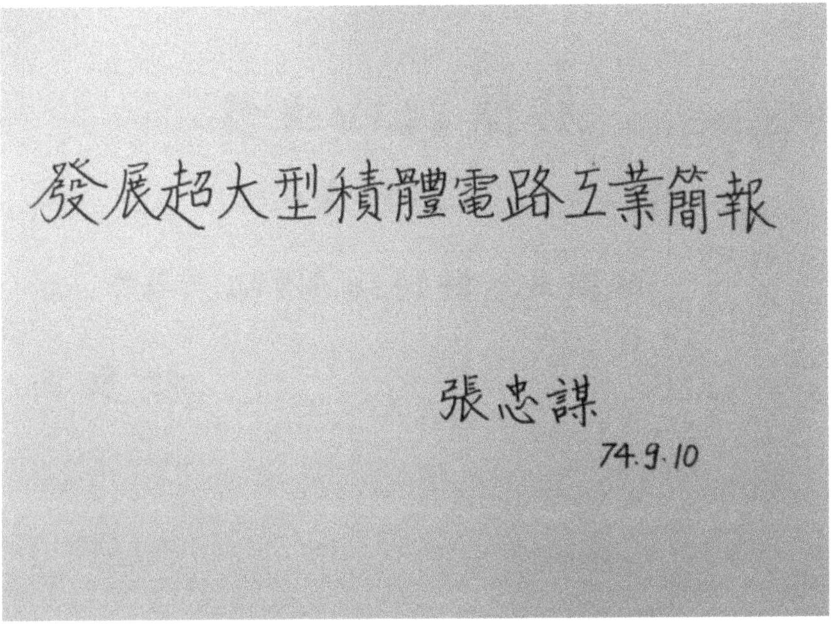

Fig. 7.1 Morris Chang's "Presentation on Developing Very Large Scale Integration Industry (發展超大型積體電路工業簡報" on September 10, 1985 ("74" is the year the Republic of China, which was established in 1911). Document provided by Ding-Hua Hu (photo by author, March 9, 2011)

what manufacturing is, but ASML, a dominant supplier of lithography systems, shows that even a simplified microchip-production sequence includes deposition, photoresist coating, lithography, etching, ion implantation, and packaging, each of which can be further broken down and involves substantial complexity, expensive equipment, and specialized services and support.

He indicates that when TSMC entered the market, Taiwan's semiconductor technology lagged 2.5 generations behind the most advanced one.[70] It is evident that the success of Taiwan's microelectronics industry is not solely related to the (most advanced) technology itself—though good technology is still necessary. There are many other factors and

[70] Chang, *Autobiography*.

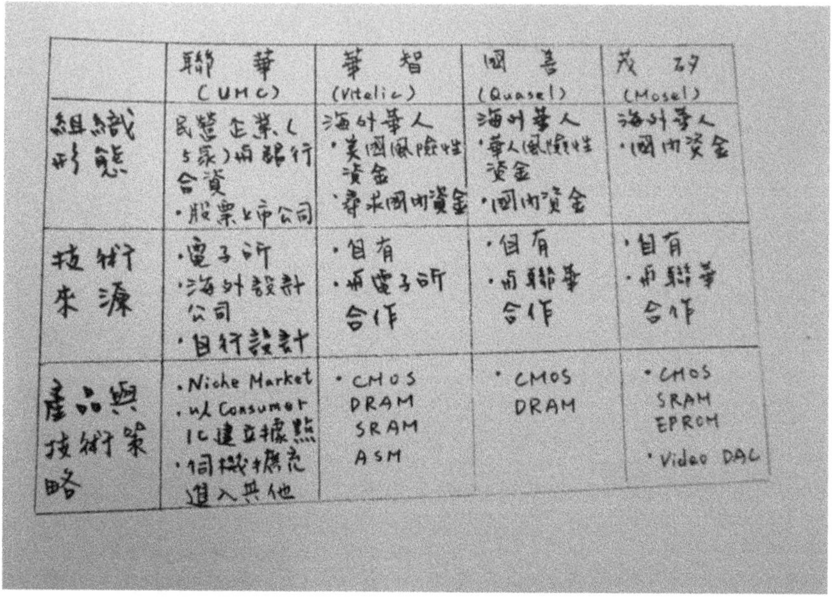

	聯華 (UMC)	華智 (Vitelic)	國善 (Quasel)	茂矽 (Mosel)
組織形態	民營企業(5家)與銀行合資 ·股票上市公司	海外華人 ·美國風險性資金 ·將來國內資金	海外華人 ·華人風險性資金 ·國內資金	海外華人 ·國內資金
技術來源	·電子所 ·海外設計公司 ·自行設計	·自有 ·與電子所合作	·自有 ·與聯華合作	·自有 ·與聯華合作
產品與技術策略	·Niche Market ·以Consumer IC建立據點 ·伺機擴充進入其他	·CMOS DRAM SRAM ASM	·CMOS DRAM	·CMOS SRAM EPROM ·Video DAC

Fig. 7.2 A Comparative Table in Morris Chang's "Presentation on Developing Very Large Scale Integration Industry" on September 10, 1985. It compares four IC companies, UMC, Vitelic, Quasel, Mosel, in terms of their organization types, sources of technology, and products and technology strategies. Document provided by Ding-Hua Hu (photo by author, March 9, 2011)

innovative ideas on about extending and transforming what IC production could be that have contributed to Taiwan's industrial dominance today.

In addition to Chang, another big figure in Taiwan's IC industry, UMC's Robert Tsao also contributed to the industry with several innovative models, including a membership foundry model and an innovative stock-bonus mechanism for employees that facilitated and shaped Taiwan's IC development. There was a priority dispute among Tsao, Chang, and others about who first thought of the idea of the pure-play foundry model, yet it is clear now that while Chang may not have invented the idea of the foundry model first, he was the first one who put the model into practice and the one who continued to refine and redefine the pure-play model.[71]

Further Thoughts: From Low to High, and a New Global Desire

The initial technology transfer from RCA played a role in fostering Taiwan's development in the industry, yet it was only part of the story. A decade after the RCA project, TSMC began an industry-transforming model in 1986 and continuously improved the model to an extent that no one can catch up so far.[72] When TSMC pioneered the pure-play foundry model—later followed by UMC—it was not initially seen as a major threat or significant development by other governments or global IDMs. Even as the pure model expanded in scale and importance, industry players and governments in more advanced countries did not seem to fully grasp the long-term implications of concentrating chip production in Taiwan until the Covid pandemic.

Taiwan already conceptualized manufacturing as a more robust and high-value field than the West did. Rather than a low-skill process to export, Taiwan looked at production as a way to secure a position within the global high-tech complex. The early success of export processing zones and electronics assembly provided a solid foundation for Taiwan's economy, making it appealing to imagine the island as both a manufacturing base and an export engine for the world. In other words, Taiwan responded to advanced countries' perception of "low" status on manufacturing by redefining it as "high" locally, with a focus on the semiconductor and computer industry. The fact that Taiwan selected extremely talented young people to complete the technology transfer exemplifies its

[71] Tsao claimed that the innovation was his, as he submitted a related proposal to the government first, yet both Hu and Chintay Shih mentioned that the US scholar Carver Mead had already had the specific idea at that time. Hu also considered that Tsao's idea was different from Chang's in that it was a membership foundry, or closer to a club's service—members would pay a premium fee to join the club, could can then get exclusive foundry services, but Chang's was open to all without such a membership condition. See Hu, Chin-Tay Shih, and Tsao, all in their *Taiwan's IT Pioneers* interviews. For the stock bonus system, see Hu and Stan Shih in their Taiwan's IT Pioneers interviews.

[72] The second largest foundry player is Samsung, with a market share of only about 7.7% in 2024.

investment in manufacturing as a process that could be made high-skill and high-value, despite Western biases.

Secondly, even though outsiders tended to flatten different types of production, the inner circles of the semiconductor profession did not homogeneously follow that view. In fact, many recognized that chip production was important. As late as 1992, five years after TSMC was founded, Jerry Sanders (AMD's co-founder and then-CEO) still insisted that "real men have fabs," combining presumptions about gender with those about the importance of manufacturing.[73] Intel also maintained its own plants to produce core products, while Japan and South Korea continuously built chip facilities. Still, these actors did not fundamentally reconsider manufacturing's strategic value. They primarily valued manufacturing because their own products required it, not because they valued manufacturing in and of itself.

Fabrication or foundry is not inherently a successful or unsuccessful creature by itself. The Taiwanese case, in retrospect, reveals that manufacturing *plus* a service mindset and engineering expertise could create an astonishingly profitable model and transform the configuration of the whole IC industry. Manufacturing is reconfigured when actors invest it with new meanings and put them into practice. But without the benefit of hindsight, Taiwan's actors at the time faced an uncertain future and interpreted the semiconductor world in diverse ways—they lived and pursued novel combinations of imagining, organizing, and enabling sociotechnical forms of life. By approaching manufacturing in this innovative way, Taiwan proudly claimed that it was a high-tech-oriented country and excelled

[73] Chang and Pan Wen Yuan Foundation, *Silicon Story of Taiwan*, 218. Although AMD, in 2008, also spun off and sold its fabrication facilities and transformed itself into a fabless company. See Advanced Micro Devices, Inc., MD and Advanced Technology Investment Company of Abu Dhabi to Create New Leading-Edge Semiconductor Manufacturing Company," news release, October 7, 2008, Exhibit 99.1 to Form 8-K, (filed October 7, 2008), U.S. Securities and Exchange Commission, accessed January 25, 2026, https://www.sec.gov/Archives/edgar/data/2488/000119312508209113/dex991.htm. Those who viewed manufacturing as important include Stephen S. Cohen and John Zysman, "Why Manufacturing Matters: The Myth of the Post-Industrial Economy," *California Management Review* 29, no. 3 (Spring 1987): 9–26, and Ross Knox Bassett, *To the Digital Age: Research Labs, Start-up Companies, and the Rise of MOS Technology* (Baltimore: Johns Hopkins University Press, 2002). Also, Ling-Fei Lin (2016) argues how manufacturing matters in the computer industry, see Ling-Fei Lin. "Design Engineering or Factory Capability? Building Laptop Contract Manufacturing in Taiwan," *IEEE Annals of the History of Computing* 38, no. 2 (2016): 22–39.

more than others in chipmaking, customer-specific services, and the broader semiconductor value chain.

In reality, Taiwan's adoption of semiconductor production was nonlinear. Although it has led the market in the pure-play foundry and even total chip outputs after the early 2000s, in another significant IC production sector, DRAM, which also pushed the technological frontier.

Like the advanced wafer foundry, it also made huge investments and suffered huge losses. With the breakup of their tech partner, Japan's Elpida, Taiwan was cast out of the business, and Korea became an absolute monopoly, with early leaders in the US and Japan occupying a very small portion of the DRAM market. Also, Samsung, as a strong IDM firm that has its own products, offers foundry services to external customers. They designed and made chips for Apple's iPhone and iPad before Apple shifted to TSMC.[74] Even within the story of RCA and TSMC, different sociotechnical imaginaries and forms of interpretative flexibility are at play. What the state officials considered was mainly a big picture of national economy and international competition; Morris Chang and Robert Tsao focused on finding a professional niche, with innovative models and new interpretive ideas; and the young talents imagined a practical way to lift Taiwan's society and electronics industry, with an ethos of serving one's country through science and technology, and practical ways to learn step by step.

Across versions of these imaginaries and interpretations, a common feature is the tendency to emphasize how Taiwan perceives its own position within the global context. For example, referring to Taiwan as an "island" is never merely a geographical description; it reflects a perspective shaped by how Taiwanese actors believe others will or should perceive Taiwan. These imaginaries consistently situate Taiwan in relation to the world. In many respects, technology is framed primarily for export and international competitiveness, rather than for domestic application, citizen use, or public welfare. In this sense, Taiwanese actors imagine Taiwan as an object of global gaze, continuously positioning it within a global context and embedding it in a specific niche of the international supply chain. This

[74] For the timelines of Apple's chips and transition to TSMC, see Kumar Priyadarshi, "Why Apple Left Samsung for TSMC: The Journey of iPhone Processors," *Techovedas*, March 17, 2025, https://techovedas.com/why-apple-left-samsung-for-tsmc-the-journey-of-iphone-processors/.

national imaginary inflected how different stakeholders approached the prospect of high-tech manufacturing.

"Manufacturing Plus" remains central to Taiwan's economic landscape. But new terms and interpretations of this work acontinue to emerge through Taiwan's ongoing interactions with state and corporate actors around the world. For example, since 2001, the semiconductor industry has been described as a "Silicon Shield," and around 2020, the new socio-technical imaginary of TSMC or chipmaking is Taiwan's "Nation-Protecting Holy Mountain." (護國神山)"[75] This defensive shield or guardian mountain image was entirely absent from the thoughts of key actors in the 1980s—whether the government, talent elites, or the general public—who could not have anticipated that TSMC would grow into a global chip giant capable of shielding Taiwan from potential attacks, but it has become a popular narrative and influences the public opinions and policy decisions.[76]

Since chips are now viewed as a critical resource[77] and advanced semiconductors constitute the basis of AI or national computing power, countries controlling them possess the capability to reconfigure global power in the twenty-first century as well. Thus, potential conflicts between Taiwan and China overshadow the status of the global supply chain surrounding the new indispensable resource. With the proliferation of social media and online news, Taiwanese society has embraced this new imaginary by imagining how the world sees Taiwan. While earlier semiconductor visions followed a top-down path driven by government and elites, this twenty-first-century imaginary emerged bottom-up from international

[75] Silicon Shield was a concept raised by Craig Addison in his book *Silicon Shield: Taiwan's Protection Against Chinese Attack* (London: Fusion Press, 2001). Addison claimed that since the United States could not tolerate Kuwait, a country that owns one fourth of the world's oil output, being occupied by Iraq, it would not accept Taiwan, the country which manufactures 70% of the worldwide outsourced ICs (integrated circuits), being attacked, either. The "holy mountains" imaginary appeared around 2015 to describe how high mountains in Taiwan protect the land from destruction by typhoons. Around 2020, this metaphor was borrowed to describe TSMC and semiconductors' importance to Taiwan's national security due to the high concentration of chipmaking in Taiwan.

[76] The holy mountain metaphor or imaginary is very popular in Taiwan's media, social media, stock market information, and everyday life today. And it has become an imaginary in which other related terms and themes appear. For example, a 2025 documentary called "A Chip Odyssey," although its Chinese title 造山者 means "Mountain Builders."

[77] Miller, *Chip War*, 45.

observers and the Taiwanese public alike, popularized through social media rather than official channels.

This new imaginary has taken effect in shaping future geopolitics, industrial policies, and societal development. Recently, TSMC invested vastly in and built new fabs in the US, Japan, and Germany, driven by substantial government subsidies, customer demands, and tax pressures, and still continued to operate their facilities in China. Taiwanese society has voiced criticism that Taiwan's own government did not prevent TSMC from distributing its fabs around the world, thereby making the "guardian mountain" less protective of Taiwan due to reduced concentration of chip production domestically, although many also complain that TSMC has absorbed many talents and resources, leading to other issues.[78] In the meantime, Japan, South Korea, Europe, and the US also have their own imaginaries for Taiwan's chip production dominance—surely not about holy mountains, but about geopolitical risks, cross-border collaboration, or the resilience of their own chip supply chains and national futures. In other words, the sociotechnical imaginaries toward semiconductors differ across time and across geographic boundaries. Concerns about a Taiwan-China war and the possible breakdown of microchip supply are only one aspect of the issue; another is many countries' strong desire to possess and control what has become a new inner engine of AI.

Conclusion: Excavating Manufacturing's Hidden Dimensions

Imaginary and reality can be more closely connected than we assume: under the proper conditions and with deliberate execution, an imagined future can become self-fulfilling. During the first two decades of Taiwan's semiconductor development, key plans and actions did not occur in a vacuum. Beyond policy incentives, technology transfer, and the creation of the dedicated-foundry model, Taiwan and TSMC's "foundregenesis" was cultivated by the sociotechnical imaginaries gradually constructed by and shared among overseas Chinese, foreign industry experts, technocratic officials, and talented engineers, each of whom represented complex social perspectives that went far beyond a set of technologies.

[78] Stella Robertson, "TSMC and Taiwan's Government: Two Boats on the Same Tide," *Domino Theory*, August 26, 2024.

Central to these desirable futures was the decision to position Taiwan in high-tech manufacturing—a sector gradually marginalized by advanced economies pursuing post-industrial society visions a half-century ago. Taiwan concentrated on becoming a global chipmaking partner, an export machine whose strategy meshed with its engineering talent pool, societal needs, and subtle political status. Together these factors laid the groundwork for TSMC's and Taiwan's remarkable rise in the global semiconductor industry over the next few decades.

In this study, I also argue for a more nuanced picture of the relationship between the social and the technological by taking both SCOT and sociotechnical imaginaries into account. In my interpretation, SCOT tends to elaborate on a relatively microscopic level and existing competing technology, whereas sociotechnical imaginaries operate on a more macroscopic scale and are more future-oriented. Combining the two therefore offers a promising approach for understanding the development of *manufacturing plus* in Taiwan. In the early stage, when Taiwan entered the semiconductor arena, both industry and technology were evolving without a clear direction and were open to many possibilities. Together with state and foreign and overseas experts, young elites followed the state's direction, but also envisioned and solidified what belongs to their versions of sociotechnical imaginaries.

Yet, once the big picture had settled, using and extending SCOT to analyze more industry-level development fits even better, as within the global IC ecology and business model, there was once a dispute with interpretative flexibility on what IC manufacturing could be, but it was finally stabilized to Morris Chang's pure-play model. In this sense of SCOT, the process can be called the social construction of manufacturing (SCOM). As I have shown in this example, companies like TSMC manufacture far more than integrated circuits—they also manufacture new models, new industrial ecology, and new social meanings of manufacturing itself, along with ever-evolving global, national, and sociotechnical imaginaries.

Bibliography

I. English

Addison, Craig. *Silicon Shield: Taiwan's Protection Against Chinese Attack.* London: Fusion Press, 2001.

Bassett, Ross Knox. *To the Digital Age: Research Labs, Start-up Companies, and the Rise of MOS Technology.* Baltimore: Johns Hopkins University Press, 2002.

Bell, Daniel. *The Coming of Post-Industrial Society: A Venture in Social Forecasting.* New York: Basic Books, 1973.

Bijker, Wiebe E., Thomas P. Hughes, and Trevor Pinch, eds. *The Social Construction of Technological Systems.* Cambridge, MA: MIT Press, 1987.

Castells, Manuel. *The Rise of the Network Society.* 2nd ed. Vol. 1 *The Information Age: Economy, Society, and Culture.* Chichester, UK: Wiley-Blackwell, 2010. Originally published 1996.

Chang, Chun-Yen (張俊彥). 2011. *Taiwanese IT Pioneers: Chun-Yen Chang*, interview by Craig Addison, February 16, 2011, Taiwan, Computer History Museum, 2012, catalog no. 102746006. https://archive.computerhistory.org/resources/access/text/2012/02/102746006-05-01-acc.pdf

Chang, Morris. 2007. *Oral History Interview*, for SEMI and Computer History Museum, https://www.semi.org/en/Oral-History-Interview-Morris-Chang

Chang, Morris. 2007. *Oral History of Morris Chang* (Chinese: 《張忠謀口述史》). Interview by Allen Patterson. August 24, 2007, Taiwan. Computer History Museum, Mountain View, CA, 2007, https://archive.computerhistory.org/resources/text/Oral_History/Chang_Morris/Chang_Morris_1.oral_history.2007.102658129.pdf

Chu, Ming-Chin Monique. "China's Defence Semiconductor Industrial Base in an Age of Globalisation: Cross-Strait Dynamics and Regional Security Implications." *Journal of Strategic Studies* 47, no. 5 (2024): 643–668. doi.org/https://doi.org/10.1080/01402390.2023.2164852.

Cohen, Stephen S., and John Zysman. "Why Manufacturing Matters: The Myth of the Post-Industrial Economy." *California Management Review* 29, no. 3(1987): 9–26.

Congress of the U.S., Office of Technology Assessment. *Technology Transfer to China.* Washington, D.C.: Congress of the U.S., 1987.

Edwards, Paul N. *The Closed World: Computers and the Politics of Discourse in Cold War America.* Cambridge, MA: MIT Press, 1996.

Hu, Ding-Hua (胡定華). 2011. *Taiwanese IT Pioneer: Ding-Hua Hu.* Interview by Ling-Fei Lin. March 9, 2011. Computer History Museum, 2012. Catalog no. 102745997. https://archive.computerhistory.org/resources/access/text/2012/02/102745997-05-01-acc.pdf

Jasanoff, Sheila. "Future Imperfect: Science, Technology, and the Imaginations of Modernity," in *Dreamscapes of Modernity: Sociotechnical Imaginaries and the Fabrication of Power*, ed. Sheila Jasanoff and Sang-Hyun Kim, 1–33. Chicago: University of Chicago Press, 2015.

Jasanoff, Sheila, and Sang-Hyun Kim. "Containing the Atom: Sociotechnical Imaginaries and Nuclear Power in the United States and South Korea." *Minerva* 47, no. 2 (2009): 119–146.

Jasanoff, Sheila, and Sang-Hyun Kim, eds. *Dreamscapes of Modernity: Sociotechnical Imaginaries and the Fabrication of Power.* Chicago: University of Chicago Press, 2015.

Kline, Ronald R. *Consumers in the Country: Technology and Social Change in Rural America.* Baltimore: Johns Hopkins University Press, 2000.

Kline, Ronald, and Trevor Pinch. "Users as Agents of Technological Change: The Social Construction of the Automobile in the Rural United States." Technology and Culture 37 (1996): 763–795.

Lécuyer, Christophe. *Making Silicon Valley: Innovation and the Growth of High Tech, 1930–1970.* Cambridge, MA: MIT Press, 2005.

Leslie, Stuart W. *The Cold War and American Science.* New York and Chichester, West Sussex: Columbia University Press, 1993.

Lin, Ling-Fei. "Design Engineering or Factory Capability? Building Laptop Contract Manufacturing in Taiwan." *IEEE Annals of the History of Computing* 38, no. 2 (2016): 22–39.

Leonard, Andrew. "Paranoid, Are Ya?" Wired, December 1, 1998. https://www.wired.com/1998/12/paranoid-are-ya/.

Marcus, George E., *Technoscientific Imaginaries: Conversations, Profiles, and Memoirs.* Late Editions, Vol. 2. Chicago: University of Chicago Press, 1995.

Mathews, John A., and Dong-Sung Cho. *Tiger Technology: The Creation of a Semiconductor Industry in East Asia.* Cambridge, UK: Cambridge University Press, 2000.

Miller, Chris. *Chip War: The Fight for the World's Most Critical Technology.* New York: Scribner, 2022.

National Advisory Committee on Semiconductors. *A National Strategy for Semiconductors: An Agenda for the President, the Congress, and the Industry.* Washington, D.C.: U.S. Government Printing Office, 1992.

Office of Technology Assessment. *Congressional Board of the 100th Congress.* Washington, D.C.: U.S. Congress, 1988.

Okimoto, Daniel I., Takuo Sugano, and Franklin B. Weinstein, eds. *Competitive Edge: The Semiconductor Industry in the U.S. and Japan.* Stanford: Stanford University Press, 1984.

Peng, Bin-Win, Houn-Gee Chen, and Bou-Wen Lin. "A Taiwan Research Institute as a Technology Business Incubator: ITRI and Its Spin-Offs." *Comparative Technology Transfer and Society* 4, no. 1 (2006): 1–21.

Pinch, Trevor, and Wiebe Bijker. "The Social Construction of Facts and Artefacts: Or, How the Sociology of Science and the Sociology of Technology Might Benefit Each Other." In *The Social Construction of Technological Systems*, edited by Wiebe E. Bijker, Thomas P. Hughes, and Trevor Pinch. Cambridge, MA: MIT Press, 1987.

Priyadarshi, Kumar. 2025. Why Apple Left Samsung for TSMC: The Journey of iPhone Processors. *Techovedas*. https://techovedas.com/why-apple-left-samsung-for-tsmc-the-journey-of-iphone-processors/. Accessed 17 March 2025.

Robertson, Stella. 2024. TSMC and Taiwan's Government: Two Boats on the Same Tide. *Domino Theory*. https://dominotheory.com/tsmc-and-taiwans-government-two-boats-on-the-same-tide/. Accessed 26 August 2024.

Shih, Chin-Tay (史欽泰). 2011. *Taiwanese IT Pioneer: Chin-Tay Shih*. Interview by Ling-Fei Lin. February 21, 2011. Computer History Museum, Mountain View, CA, 2012. Catalog no. 102745999. https://archive.computerhistory.org/resources/access/text/2012/02/102745999-05-01-acc.pdf

Shih, Stan (Chen-Jung) (施振榮). 2012. *Taiwanese IT Pioneer: Stan (Chen-Jung) Shih*. Interview by Ling-Fei Lin. March 21, 2011. Computer History Museum, 2012. Catalog no. 102746001. https://archive.computerhistory.org/resources/access/text/2012/02/102746001-05-01-acc.pdf

Taiwan Semiconductor Manufacturing Company (TSMC). 2006. *2025 Annual Report*. Hsinchu, Taiwan: TSMC.

Taiwan Semiconductor Manufacturing Company (TSMC). 2021. *2020 Annual Report*. Hsinchu, Taiwan: TSMC.

Taiwan Semiconductor Manufacturing Company (TSMC). 2025. *2024 Annual Report*. Hsinchu, Taiwan: TSMC.

Taylor, Charles. *Modern Social Imaginaries*. Durham, NC: Duke University Press, 2004.

Toffler, Alvin. *The Third Wave*. New York: William Morrow, 1980.

Tsao, Robert H. C. (曹興誠). 2011. *Taiwanese IT Pioneer: Robert H. C. Tsao*. Interview by Ling-Fei Lin. February 17, 2011. Computer History Museum, 2012. Catalog no. 102746012. https://archive.computerhistory.org/resources/access/text/2012/02/102746012-05-01-acc.pdf

U.S. Department of Commerce, Industry and Trade Administration, Office of Producer Goods. *A Report on the U.S. Semiconductor Industry*. Washington, D.C.: U.S. Government Printing Office, 1979.

Yang, D. Y. (Ding-Yuan) (楊丁元). 2012. *Taiwanese IT Pioneer: D. Y. (Ding-Yuan) Yang*. Interview by Ling-Fei Lin. February 23, 2011. Computer History Museum, 2012. Catalog no. 102745994. https://archive.computerhistory.org/resources/access/text/2012/02/102745994-05-01-acc.pdf

II. Chinese (English titles translated directly from Chinese ones)

Chang, Chun-Yen (張俊彥), and Po-Long Yu (游伯龍), eds. *Vitality: How Taiwan Creates the Miracles of Semiconductor and Personal Computer Industries* (Chinese:《活力-台灣如何創造半導體與個人電腦產業奇蹟》). Taipei: China Times, 2002.

Chang, Ju-Shin, and Pan Wen Yuan Foundation (張如心/潘文淵文教基金會). *Silicon Story of Taiwan* (Chinese《矽說台灣》). Taipei: CommonWealth Publishing Group, 2006.

Chang, Kuo-li (張國立). *Biography of Sun Yun-Suan* (Chinese: 孫運璿傳). Taipei: Commonwealth Publishing, 2010.

Chang, Morris (張忠謀). *Autobiography of Morris C. M. Chang, Vol. 1: 1931–1964* (Chinese: 《張忠謀自傳 上冊:1931–1964》). Taipei: CommonWealth Publishing, 1998.

Hong, Yi-Yan (洪懿妍). *Innovation Engine: ITRI, The Helper of the Success of Taiwan Industry* (Chinese:《創新引擎:工研院:臺灣產業成功的推手》). Taipei: CommonWealth Magazine, 2003.

Lin, Ling-Fei (林玲妃). 1996. Taiwan's IC Industry, Standing at the Crossroads (Chinese:《處於十字路口的台灣IC工業》). *China Times*, September 26, 1996.

Liu, Jinqing (劉進慶). *An Economic Analysis of Taiwan After the War* (Chinese:《台灣戰後經濟分析》). Taipei: Ren-Jian Publishing Company, 1992.

Scott, Maurice. 1994. Taiwan's Trade Development (Chinese: 「臺灣的貿易發展」). Trans. by Chao-Wu Lin. In *The Paper Selections of Foreign Trade Development in Taiwan* (Continuation) (Chinese: 《臺灣對外貿易發展論文集(續編)》), ed. by Chi Schive, 6–35. Taipei: Linking Books.

Shih, Chin-Tay (史欽泰), Chen Tien-Chih (陳添枝), and Wu Shu-Min (吳淑敏). 2025. *From the Periphery to the Core: How Taiwan's Semiconductor Industry Became the Heart of the World* (Chinese: 《從邊緣到核心:台灣半導體如何成為世界的心臟》), Kindle ed.

CHAPTER 8

Teaching the History of Emerging Technologies in an Information Age

Honghong Tinn

Introduction

This chapter proposes that historians of technology are well equipped to foster critical discussions about urgent issues for the information age in which we live. Historians of technology, with their rich understanding of past debates, are well-positioned to investigate how new technologies have been developed, introduced, promoted, and experienced by individuals or organizations. Since the 1980s, the percentage of undergraduate history majors has been dropping in the United States. The number for 2019, 1.2%, was the lowest since 1949.[1] But today's obsession with technology offers plenty of opportunities for historians of technology to

[1] Robert B. Townsend, "Has the Decline in History Majors Hit Bottom?" *Perspectives on History*, February 23, 2021, https://www.historians.org/perspectives-article/

H. Tinn (✉)
University of Minnesota Twin Cities, Minneapolis, MN, USA
e-mail: htinn@umn.edu

J. L. Lieberman, A. Mohsin (eds.), *Technological Humanities*,
https://doi.org/10.1007/978-3-032-24741-4_8

provide expertise and insights—particularly in the classroom. Our unique strength is our insistence on understanding, documenting, and analyzing technology's technical details and the social context of any technology's construction. Our attention to the technical aspect of a technology distinguishes us from other humanities and social sciences scholars. We are familiar with the hype that is associated with new features of emerging technologies, decisions regarding employment and unemployment that accompany the rise of technologies, and politics and inequalities that have evolved alongside seemingly promising technologies. We have convincing stories to share with students as they reflect critically on seemingly trendy and powerful technologies.

In this chapter, I discuss pedagogical insights historians of technology can offer in the age of artificial intelligence (AI), algorithms, and computer-fueled automation, drawing from my experience teaching a lower-division course titled "Digital World" at the University of Minnesota, Twin Cities. I share with readers the design of my course as a historiographical review of studies about the internet and social media. Some of my notes may resonate with other scholars who want to help their students examine emerging technology critically and reflectively.

Historians of technology have contributed to debates on "the information age," a concept former president of the Society for the History of Technology (SHOT), Ronald R. Kline, has historicized. Years before he published *The Cybernetics Moment: Or Why We Call Our Age the Information Age,* he explored these themes in the classroom in an undergraduate course titled "Inventing an Information Society" at Cornell University.[2] In both his teaching and his published research, Kline has chronicled the rise, expansion, and fracturing of the discipline of cybernetics, anddemonstrated that the concepts "information" and "infor-

has-the-decline-in-history-majors-hit-bottom-data-from-2018-19-show-lowest-number-since-1980-march-2021/

[2] Ronald R. Kline, *The Cybernetics Moment: Or Why We Call Our Age the Information Age* (Baltimore, MD: Johns Hopkins University Press, 2015). See also Kline, "Cybernetics, Management Science, and Technology Policy: The Emergence of 'Information Technology' as a Keyword, 1948-1985," *Technology and Culture* vol. 47 no. 3 (2006): 513–535; Kline, "Cybernetic as a Useable Past, Review of Andrew Pickering: *The Cybernetic Brain: Sketches of Another Future*," *Metascience* 20 (2011): 519–524; Greg Adamson, Ronald R. Kline, Katina Michael, and M. G. Michael, "Wiener's Cybernetics Legacy and the Growing Need for the Interdisciplinary Approach [Scanning Our Past]," *Proceedings of the IEEE* 103, no. 11 (2015): 2208–2214.

mation age" were consistently re-interpreted by both academics and critics over time. The central issue that new technology poses to a society, according to Kline, is the relationship between humans and machines. In his Leonardo da Vinci Medal address at the 2016 SHOT meeting he focused on two themes in that relationship: "the use of technology; and the attempts to create a science that could explain what it meant to be human in an age of intelligent machines."[3] In this context, it can be productive for scholars, critics, and the public to reject the view that the information age is an era that demands we accept its new terms. Instead, we should carefully investigate how the new era is constructed. Historians of technology, as Kline's work exemplifies, are well-suited to encourage college students to engage critically with analyses of technologies and work with students to evaluate the roles different stakeholders play in shaping the future of these technologies.

Dr. Thomas J. Misa, also a former president of SHOT, created the Digital World course at the University of Minnesota. The course fulfills general education requirements. Misa included a series of fascinating themes in the course: the structure of the internet; compressing reality; security and surveillance; virtual reality; and the future of the internet. These themes covered both historical analysis and contemporary cases, with a focus on the increased technical capacity of the internet to surveil and censor users. I began teaching this course in spring 2020, after I joined the program in the history of science, technology, and medicine at the University of Minnesota. I have an average of 160 undergraduate students each spring semester. In designing my own version of the course, I drew from Misa's syllabus, as well as Kline's "Inventing an Information Age" course, for which I served as a teaching assistant. Currently, I structure the course with the following themes: the history of the ARPANET (Advanced Research Projects Agency Network) and the internet, the Arab Spring, social media and monitory democracy, the Cambridge Analytica-Facebook data scandal, and ethical issues surrounding facial recognition systems, algorithms, the gig economy, and artificial intelligence. I encourage students to think of the development of internet and internet-enabled

[3] See p. 835 in Kline, "Da Vinci Medal Address: Humans and Machines," *Technology and Culture* 58, no. 3 (2017): 835–845. The Leonardo da Vinci Medal address took place at the annual meeting of SHOT in 2016, and the address was published the year after. The Leonardo da Vinci Medal is the highest recognition from the Society for the History of Technology to acknowledge an individual's outstanding contribution to this scholarly field.

technologies as a process shaped by contingent historical factors instead of following a predictable trajectory. I also expect them to become critical thinkers and careful interpreters of texts and everyday narratives of science, technology, and society. Throughout the semester, they learn to evaluate the roles different stakeholders play within society in shaping the future of internet-enabled technologies.

In this chapter, I share the design of my course, including assigned readings and essay questions that students work on. My course design focuses on three "toolkits" that historians of technology use. These include: (1) using social groups as an instrumental classroom framework, (2) identifying counternarratives, and (3) introducing cybernetics as a useable past. Specifically, first, I focus on the instrumental concept of "social groups" in teaching the history of technology. I draw from the school of social construction of technology's (SCOT) emphasis on "social groups," which Kline also elaborated in his work.[4] The identification of "social groups" can encourage students to keep an open mind about the varied historical trajectories a technology can take. Thus, they may be willing to consider a variety of solutions to remedy the problems related to the uses and misuses of new digital technologies today. Second, I discuss the concept of "counternarratives," as Kline elaborated in his 2012 SHOT presidential address, in light of Nye's *America as Second Creation*.[5] Counternarratives compete with mainstream narratives about new technology told by corporations. The identification of counternarratives characterizes the unique storytelling that historians of technology can deliver. Third, I share how Kline's research on cybernetics helped me emphasize metaphors and automation in the construction of an information age. The toolkits help me and my students discuss the debates generated during the age of artificial intelligence, algorithms, and computer-fueled automation.

[4] For the concept of social groups, see, for example, Trevor J. Pinch and Wiebe Bijker, "The Social Construction of Facts and Artifacts: Or, How the Sociology of Science and the Sociology of Technology Might Benefit Each Other," in *The Social Construction of Technological Systems*, ed. Wiebe Bijker, Thomas P. Hughes, and Trevor J. Pinch (Cambridge, MA: MIT Press, 1987 [1984]), 17–50; Ronald R. Kline and Trevor J. Pinch, "Users as Agents of Technological Change: The Social Construction of the Automobile in the Rural United States," *Technology and Culture* 37, no. 4 (1996): 763–795; Ronald R. Kline, *Consumers in the Country: Technology and Social Change in Rural America* (Baltimore, MD: Johns Hopkins University Press, 2000).

[5] David E. Nye, *America as Second Creation* (Cambridge, MA: MIT Press, 2003).

Below, I will discuss both the scholarly works assigned for this course and how my students bring these texts together. The history covered during the first half of the semester spans ARPANET's origins to its retirement in 1990, a shift from military funding supporting the research and development of computer networks to for-profit corporations taking over the management of computer networks. I emphasize social groups and counternarratives in the first half of the course. For the second half of the semester, I focus on a period from the mid-1980s to the present day. Early "virtual communities" on bulletin board systems emerged in the mid-1980s. Those communities share similarities with today's social media. The analytical themes that students work on include the gradual appearance of characteristics such as surveillance and corporate interests embedded in recent internet-based technologies. I design three modules for this period: democracy, surveillance, and automation. I focus on counternarratives in this part of the course. Moreover, Kline's research on cybernetics, especially his discussion about metaphors of human brains and the mind, and his interest in Norbert Wiener's concerns over automation, have helped me guide students to understand the assigned scholarly works.

Learning from the History of the ARPANET and the Internet

For the first week of the semester I assign E.M. Forster's 1909 science fiction story "The Machine Stops." The short story allows students to consider that the governance of the internet can affect or even restrict what people would like to do, which is different from a much more popular belief by many college students that the internet provides a space for everyone to express their thoughts online. Forster's short story delineates a dystopian future featuring an internet-like communications network.[6] In this dystopia, where few people are interested in in-person visits, machine intelligence constructs a world for human beings to live underground and interact with each other through conferencing systems. After a protagonist escapes from the underground world, machine intelligence prohibits human beings from going to the surface of the Earth. The short story encourages students to ask how a mundane or ubiquitous technology can

[6] E.M. Forster, "The Machine Stops," 1909, https://www.gutenberg.org/cache/epub/72890/pg72890.txt

be designed, regulated, or controlled by an invisible entity with specific goals. It serves as a prompt for students to begin to deliberate over Melvin Kranzberg's thesis, "technology is neither good nor bad; nor is it neutral."[7]

We then move from that fictional short story of 1909 to the history of the Cold War and the building of the ARPANET, the predecessor of the internet, by reading several acclaimed works by historians of technology: Janet Abbate's book *Inventing the Internet* on the history of the ARPANET in the United States, Fred Turner's article on *The Whole Earth Calatog* and the Whole Earth 'lectronic Link (WELL), and Eden Medina's article on Project Cybersyn, an early computer network implemented in Chile in 1971.[8]

Abbate's book chapters chronicle the groups and organizations that worked together or independently to build the ARPANET from the 1960s to the 1980s. While military funding and applications prompted the construction of the first computer networks in the United States, ARPANET's civilian designers and users across United States university campuses also contributed to the completion, operations, and improvement of ARPANET. Abbate discusses how National Science Foundation (NSF) funding expanded the technical infrastructure of computer networks beyond ARPANET, hence strengthening the growth of the "internetworking" of computer networks, especially in the 1980s. After ARPANET was terminated in 1990, and the NSF's backbone system was terminated in 1995, the era of "US government ownership of the internet's infrastructure" ended. The internet thus became an arena for for-profit corporations.[9] Fred Turner's work offers a different perspective to consider: social cultures promoted the growth of early virtual communi-

[7] Melvin Kranzberg, "Technology and History: 'Kranzberg's Laws,'" *Technology and Culture* 27, no. 3 (1986): 544–560.

[8] Janet Abbate, *Inventing the Internet* (Cambridge, MA: MIT Press, 2000); Fred Turner, "Where the Counterculture Met the New Economy: The WELL and the Origins of Virtual Community," *Technology and Culture* 46, no. 3 (2005): 485–512; Eden Medina, "Designing Freedom, Regulating a Nation: Socialist Cybernetics in Allende's Chile," *Journal of Latin American Studies* 38, no. 3 (2006): 571–606. Turner and Medina also published monographs, see Turner, *From Counterculture to Cyberculture: Stewart Brand, the Whole Earth Network, and the Rise of Digital Utopianism* (Chicago: University of Chicago Press, 2006); Eden Medina, *Cybernetic Revolutionaries: Technology and Politics in Allende's Chile* (Cambridge, MA:
MIT Press, 2011).

[9] Abbate, *Inventing the Internet*, 199.

ties in the 1980s. This perspective contrasts to the military funding and applications delineated in Abbate's work. Turner focuses on the role of New Communalism in the late 1960s in stimulating the emergence of one of the early virtual communities, 1985's Whole Earth 'lectronic Link (WELL). Specifically, through Turner's article, students investigate *The Whole Earth Catalog*'s attention to small-scale technologies and the *Catalog*'s positive interpretation of computers and calculators, a stance that contrasts to the New Left's criticism of the military-industrial-academic complex. Usually, students do not anticipate how ARPANET or followers of New Communalism can relate to their media environments. By the time we finish Abbate's and Turner's writing, they can acknowledge how, over time, different groups of people and institutions exercised their influence to permit or restrict who were allowed to use early computer networks for what. As such, they can acknowledge that the internet today is also not free from influences interest parties attempt to exercise.

Students then read Eden Medina's work to compare a Chilean computer network with the ARPANET. The Cybersyn project in Chile that connected several computers as a network was short-lived, lasting from 1971 to 1973, but it was an ambitious project—its democratic socialist designers aimed to use "computer and communications technology to bring about structural change in society."[10] Specifically, President Salvador Allende and his technocrats sought a decentralized computer network to materialize a "decentralised governance with worker participation in management, reinforcing his professed belief in individual freedoms."[11] Through this case, students deliberate on whether they agree with designers' presumption that the computer network could transform Chilean society to democratic socialism. As Medina sophisticatedly demonstrates, some Chilean engineers and workers doubted the technology could directly trigger an ideological change. By simply using materials provided by Medina in the reading, students run "reality checks" about the hurdles encountered by policy supporters at the implementation level, which are contrary to the idealistic goals the Cybersyn project intended to achieve. As this case reveals the political agenda embedded in a computer network, students further consider if they see similarities with the case of ARPANET, which emerged from the US military-industrial-academic complex.

[10] Medina, *Cybernetic Revolutionaries*, 6.
[11] Medina, "Designing Freedom," 596.

At the end of the first half of the semester, students watch the documentary *Digital Disconnect: How Capitalism is Turning the Internet against Democracy, which introduces arguments* Robert W. McChesney developed in his book of the same title.[12] He argues that we need not pay attention to the debate on whether the internet has the potential to promote democracy or not. Instead, capitalism is the ultimate underlying power that dominates both the internet and democracy. Therefore, we should pay more attention to reveal how capitalist mechanisms undermine democracy and how large profit-driven corporations get to decide what the internet can offer. *Digital Disconnect* guides students to distinguish correlation (between the internet and democracy) and causation (between capitalism and democracy). When reasoning in historical and other qualitative research in social sciences and humanities, scholars have been using shark attacks and ice cream consumption as an analogy to explain mistaking correlation for causation. During the summer months the occurrence of shark attacks often coincides with a rise in ice cream consumption. Yet, one does not cause the other: the factor that affects both ice cream consumption and shark attacks is the summer. I often use cartoon images of shark and ice cream to attract students' attention, and find them interested in the analogy and McChesney's argument.

The documentary is informative, persuasive, and well-made with rich images and news video clips, hence a perfect piece to vary the format of course materials. As many students are frequent users of social media, they tend to assume social media is an effective tool for social changes. But, with McChesney's documentary and the assigned readings of the previous weeks, students are exposed to scholarly analyses of other factors that influence or even restrict what social media can offer. In considering McChesney's argument and evidence in their discussion sections and written assignments, students more willingly reconsider their assumptions about the role of social media in society.

[12] Robert W. McChesney, *Digital Disconnect: How Capitalism is Turning the Internet against Democracy (New York:* The New Press, 2013). Documentary *Digital Disconnect: How Capitalism is Turning the Internet against Democracy,* directed by Jeremy Earp, 2018.

Social Groups and Politics in the History of Technology

In my lectures and discussion sections, I especially emphasize the role of social groups. In the pieces by Abbate, Turner, Medina, McChesney, and Forster, students and I review the designers, promoters, opponents, and users of computer networks, the characteristics of each computer networks, and the role of computer networks in each society or historical temporality. Students are now ready to work on the following essay prompt—choose and articulate three of the most important organizations, groups of people, or factors that significantly shaped the history of ARPANET and the internet. They can discuss, for example, military funding, military applications, ARPANET's civilian designers and early users, NSF funding, New Communalism, and commercial interests. The prompt is one of two essay prompts that students may choose to write about in concluding the first half of the semester.

For the second essay option, I ask students to consider social groups differently. With the works of Abbate, Turner, Medina, McChesney, and E.M. Forster, students are now ready to work on answering the following question—"Do computer networks always promote decentralized communication?" To answer this question, I prompt "to compare and contrast the histories written by Abbate, Turner, and Medina, as well as the short story the Machine Stops and McChesney's documentary."[13] First, this assignment asks students to consider technological determinism and social determinism (or social construction of technology). Students may review each scholar's definition of the term "decentralized." They may also review historical actors' definitions of the term "decentralized" or students can define what they consider "decentralized." For example, they may associate the term with anti-hierarchical, democratic, or distributed practices. Second, after learning from discussion sections and lectures, students are familiar with integrating social groups in their analysis at the point of working on this assignment. By paying attention to social groups, students can better discuss their theses with concrete historical examples in their essays. Assignments like this one help students put their own voices in

[13] Students only need to select three readings to answer this question. Students may take either side, and the grading is based on the quality of their analysis of the selected three readings, rather than the side they take.

conversation with historians of technology—a practice that promotes their critical thinking of present-day technologies as well as historical ones.

My selection of assigned readings and my emphasis on social groups have been inspired by the school of social construction of technology, especially the writings of Trevor J. Pinch, Wiebe Bijker, Kline, and others. In this course, I emphasize what SCOT means by social groups, including designers, scientists, engineers, users, activists, and others, including mediators, who shape technological choices in science and technology.[14] In my view, the term "social groups" can also refer to stakeholders. By emphasizing social groups, I invite students to take an approach that considers a full range of actors and contexts when reviewing and analyzing technology and its social world in a historical temporality. Identifying as many "social groups" as possible and articulating their goals, views, and cultures works as a classroom strategy that encourages students to read closely both primary and secondary sources. Discussion section leaders and I often include in a PowerPoint slide a list of social groups and ask students to examine sources to note what each group advocates. Students take turns to present historical roles they have researched from the reading, thereby introducing a variety of perspectives into the class. For example, we often host a stakeholder discussion about Medina's article, in which we ask students to present the viewpoints of cyberneticians, technocrats, factory workers, and other individuals or groups based on the information included in the article. Just the process of identifying these social groups can complicate deterministic narratives, which imply that a technology, such as ARPANET or artificial intelligence, is an autonomous actor, and would revolutionize everything. Students who learn about SCOT also learn that a technology does not function in this deterministic manner. Certain social groups may invest in a narrative about determinism; but other social groups can promote counternarratives that highlight the role human choices play in adoption and use.

Kline has discussed the role of social groups in the history discipline in his 2016 Da Vinci Medal address. The address elaborates on the use of technology as an aspect of the historical relationship between humans and machines.[15] Kline recalls that "the only major point" on which Pinch and he disagreed when working together on the classic paper "Users as Agents of Technological Change" was the respective scope of social groups they

[14] Please see footnote 4.

[15] Kline, "Humans and Machines," 835.

would like to include in the paper. Pinch emphasized the interpretations of a specific set of social groups that prompt changes of an artifact. Kline was interested in including interpretations from all types of social groups. I found the following excerpt about their different foci further illustrates the value of "social groups" in classroom for learning about the history of technology.

> Trevor wanted to discuss only those interpretations of the car that led to changes in an artifact. For example, using the car to plow the fields prompted companies to bring out tractor kits for autos, and using the car to do the wash prompted manufacturers to add a power-takeoff option to tractors. Trevor wanted to focus on the theme of users as agents of technological change in order to retain the sharp analytical edge of SCOT. I wanted to talk about all social interpretations of the rural car, uses that led to technological changes and those that did not, such as farm women using the car to run their egg business and expand their traditional gender roles. In the terms of philosopher Isaiah Berlin, it was a classic clash between the hedgehog who saw the world through one grand idea and the fox who wanted to explore many things.[16]

While co-authoring an article, Pinch and Kline had to decide on whether to pursue theoretical integrity or a broad survey of all interpretations existing in a historical temporality. The "friendliest of disputes," as Kline termed, was settled as follows: the co-authored journal article did emphasize the theoretical aspect with specific interpretations, whereas Kline was able to include a broader survey of various interpretations in his book *Consumers in the Country*.[17] In undergraduate classrooms, instructors have flexibility to implement both approaches and hence allow students to explore the diverse views held by different social groups in the past. They can emphasize those stimulating changes and chronicle those failing to bring about changes. Both approaches, to me, effectively encourage students to consider the agents of social change or technological change critically.

In the context of social and technological change, "social groups" can be instrumental in teaching about emerging technologies in an information age. Keep in mind that today's undergraduate students have been born generally about ten years after the founding of Google in 1994.

[16] Kline, "Humans and Machines," 838–839.

[17] Kline, "Humans and Machines," 839 and Kline, *Consumers in the Country*.

Another benchmark is Facebook's founding year—2004. The internet and social media are a mundane technology to them. That is to say, they are native to social media. Thus, teaching the history of the internet is not only about studying why we have the internet today. It is also about equipping students with tools to explore how the internet was imagined, designed, created, managed, and used in various ways in the past. The history hopefully allows students to consider different forms, including an even more democratic form, of social media. In my experience of reading essays from students and dropping by discussion sections over time, works authored by Abbate, Medina, Turner, and McChesney, as well as my emphasis of social groups, help students better imagine the shuffling and re-shuffling of power dynamics, cultural symbols, and economic relations that can emerge after a digital technology is introduced to our society.

Identification of Counternarratives About Technology

An emphasis on "social groups" in teaching the history of the information age can be paired with an emphasis on the "counternarrative" Kline elaborated on in his 2012 SHOT presidential address.[18] In this address, Kline first discussed Rebecca Herzig's call for a consideration of "the morals" of historians of technology's storytelling, "be they progressivist or declensionist, triumphant or tragic."[19] He then noted that in David Nye's *America as Second Creation*, Nye presented both the "technological foundation narratives" constructed by white writers about technologies during the westward expansion, as well as counternarratives voiced by Native Americans, laborers, farmers, and reformers about the same technologies' destructive effects. In the following excerpt from the address, Kline identifies the counternarratives that historians of technology have successfully delivered. The counternarratives are different from mainstream narratives about technology, especially those about great minds and their inventions:

[18] Kline, "Presidential Address: Foundational Stories," *Technology and Culture* 54, no. 1 (2013): 117–129. The presidential address took place at the annual meeting of SHOT in 2012.

[19] Rebecca Herzig, "A Thing for Stories," *Technology and Culture* 50, no. 3 (2009): 613–622.

> Nye's book reminds me that we, as technology's *scholarly* storytellers, do not construct the dominant narrative told about technology in today's world. That has long been told by a host of other storytellers: corporations, politicians, journalists, advertising agencies, utopian novelists, biographers, and teachers. For the past thirty years, historians of technology have constructed a well-documented counternarrative. We have argued that technology is not reducible to applied science; that technology, on its own, does not drive history; that government promotes as much as it regulates technology; and that invention is a social process rather than the heroic act of a Thomas Edison or a Steve Jobs.[20]

Teaching about the history of what we call the information age, historians of technology consistently battle many compelling dominant narratives told by the social groups listed in the above quote. To broaden students' imagination of ways in which one can design, use, or envision technologies beyond dominant narratives, I see our role as resourceful instructors who introduce into the classroom a diverse set of social groups and their counternarratives of a historical temporality. That imagination is especially crucial for the second half of the semester of my course as I include readings about contemporary internet-enabled technologies designed and, to a great extent, controlled by large tech corporations. These readings focus on how these technologies can engender seemingly harmless but severe political consequences. By identifying counternarratives in the history of technology through the assigned readings, student may identify both the positive and negative consequences of a technology at the beginning of its adaption.

On the role of historians of technology, Kline also noted in 2012, "as Herzig says, our stories do engage the morality, the politics, of technology, whether we want them to or not. But we should not assume, as I often have, that someone else will faithfully translate those stories into political action."[21] Kline agrees that the storytelling of historians of technology is no doubt an engagement with the politics of technology. We can reasonably hope, but should not assume, that our teaching can occasionally empower our students to make some differences in the future. Thus, as instructors in colleges, historians of technology may hone skills and gain expertise to present great stories to tell in class, but we also need to acknowledge that is probably not enough to push political change.

[20] Kline, "Foundational Stories," 119.

[21] Kline, "Foundational Stories," 119.

Social Media and Democracy

For the later part of the semester, the course is centered on the internet's history since the mid-1980s, from virtual communities of bulletin board systems to social media and their algorithms. Democracy, surveillance, and automation are the three modules I implement in the second half of the semester. We discuss how, over time, these internet-based technologies have been turning into surveillance tools, exercising negative effects on democracy, and automating decisions that were originally made by human beings.

Similar to "The Machine Stops," I assign students a fictional story to begin the second half of the semester: "Nosedive" (Netflix's *Black Mirror* series, 2016). The protagonist in "Nosedive" lives in a dystopian future where everyone can rate each other on a scale from one to five stars through an interface on contact lens-like eye implants. The rating activities can be based on any personal interactions ranging from viewing each other's social media-like postings, to chatting with strangers when buying a coffee at a shop, to talking with a taxi driver. The protagonist aspires to increase her score so that she can apply for a luxurious housing property. But she overdoes it and eventually her score plunges after she can no longer be nice to everyone. The episode pushes students to compare the score to the number of "likes" and followers on social media or to their ratings as users in a gig economy. To promote this perspective, I accompany this viewing with another assigned reading, a book chapter, "The Internet of me (and my 'friends')" from a popular textbook *Misunderstanding the Internet*. The chapter deliberates whether social media offers equal opportunities to everyone to express one's voice online. As author Natalie Fenton claims, "Everyone can speak, anyone can blog, everyone is a producer – what wonderful plurality, what joyous freedoms. Yet the power endowed through such plurality and freedom may be hollow and fleeting. There is not a limitless stream of power that everyone can equally lay claim to."[22]

"Nosedive" subtly argues that the rating system and one's social-economic status are correlated. In this dystopia, the richer tend to have higher scores. A truck driver whom the protagonist meets in the episode

[22] Natalie Fenton, "The Internet of Me (and My 'Friends')," *Misunderstanding the Internet*, authored by James Curran, Natalie Fenton, Des Freedman (London: Routledge, 2016, 2nd Edition), 145–172. See p. 165.

has an extremely low score. The truck driver and her husband used to have relatively high scores. But she chose to ignore the rating system after her husband, who was suffering from cancer, could not enroll in clinics that demanded extremely high scores for experimental treatments. In other words, her husband was not morally good enough and rich enough to be qualified for certain experimental treatments. The episode illuminates Fenton's argument that social media is part of our real world, and thus it articulates the dynamics of the political and economic environments people are situated in. She notes, "far from broadening our communicational horizons and deliberative understandings, social media work to reinforce already existing social hierarchies and further strengthen close(d) communities."[23] *"Nosedive"* and "The Internet of me (and my 'friends')" encourage students to analyze critically the technology they use on a daily basis.

After exploring how *"Nosedive"* and Fenton's analysis can help explain some of their observations of social media in lectures and discussion sections, students are prepared to work on the relationship between the internet and democracy more directly. We temporarily put McChesney's analysis of the United States aside and look into cases in which states could directly utilize the internet to achieve political goals. Students read excerpts from *Misunderstanding the Internet* to learn about China's censorship on the internet, the Arab Spring, and a concept called monitory democracy—a specific type of civilian power enabled by networked mobile technologies. It is a concept developed by political scientist John Keane to characterize the potential of a media-saturated society to enhance its degree of democracy. Specifically, with some digital technologies, individuals can scrutinize or supervise politicians, governments, law enforcement, and political parties.[24] For example, protesters in Europe and Hong Kong as well as African Americans in the United States used their mobile phones to record police brutality and shared it online to advocate for themselves.

Cases in following readings then push students to think about how far the concept monitory democracy can go. Students are assigned to read excerpts from *Targeted*, authored by Brittany Kaiser, a whistleblower of the Cambridge Analytica-Facebook data scandal. Cambridge Analytica, a

[23] Fenton, "The Internet of Me," 150.

[24] John Keane, "Monitory Democracy and Media-saturated Societies," *Griffith Review* 24 (2009): 47–69.

political campaign company, used breached Facebook data to swing the vote on Brexit and the 2016 presidential election in the United States. Data of somewhere between 30 and 70 million Facebook users, who were unaware that their data were being used, were acquired by the company as early as 2014. The data were then manipulated in an unprecedented, sophisticated mechanism. Users were categorized into five types of personalities.[25] Data scientists were most interested in users who showed more "neurotic" traits, because they were regarded as persuadable. They were then microtargeted on social media for refined, personalized campaign videos, from a pool of over 10,000 iterations of each video.[26]

Teaching *Targeted* has been an interesting experience. During a semester of Zoom-based teaching due to the COVID pandemic, a student was quite upset with my criticism of the misuse of data in the Cambridge Analytica-Facebook data scandal. The student, who might be an aspiring businessperson, raised their hand in a Zoom webinar and vehemently defended the right of corporations to use data without restrictions. The student believed that corporations are all about making profit; if the data is there, why cannot they make profit from it? While I have been always prepared for disagreements from students, I have not often encountered college students who expressed fervently about their belief in deregulation and their defense to have the freedom to make profits. But it was a great experience for my next iteration of the course to emphasize why some business practices, as illustrated in assigned course materials, were unethical and can be unlawful. It has also been interesting to observe how undergraduates have gradually become unfamiliar with the Cambridge Analytica-Facebook data scandal. In the first few iterations of the course, many students enthusiastically told me that they did projects about the scandal in high school and were excited about reading *Targeted* and other course materials addressing social media and democracy. But in recent iterations, it seems that because the scandal happened almost a decade ago, the number of students who knew about the scandal before taking this course has decreased. The fading of the case among the student population demonstrates that historians play an important role in chronicling

[25] The big five personality test is also known as the OCEAN test, which identifies traits of openness, conscientiousness, extraversion, agreeableness, and neuroticism in one's personality.

[26] Brittany Kaiser, *Targeted: The Cambridge Analytica Whistleblower's Inside Story of How Big Data, Trump, and Facebook Broke Democracy and How It Can Happen Again* (New York: HarperCollins, 2019).

and periodizing algorithmic social media and their interweaving with democratic mechanisms.

After focusing on the United States, we shift to examine two additional international examples showing that states could easily utilize the internet to achieve political goals: *Al Jazeera*'s documentary *The Spy in Your Phone* (2021) and the BBC news program's *Weapons of Mass Surveillance* (2017). I assign the two videos to discuss how the spyware Evident and Pegasus became tools for authoritarian states to surveil and arrest dissidents.[27] These cases reveal an arms race of digital tools between states and their citizens. But it seems that states are on the winning side. While monitory democracy occasionally boosts the power of citizens, overall, states enjoy more resources to equip themselves with digital tools to surveil their citizens.

Another similar example that illustrates the limited power of individuals with their digital technology is about the malware industry. Fenton in her chapter points out, according to a 2012 study, that the amount of malware created has exceeded that of legitimate software.[28] Although it is difficult to estimate the exact amount of malware produced every year, we can assume a significant effort is invested in sabotaging the use of technology.

Based on the above readings, students are now well-prepared to write an essay at the end of the semester to explain how the internet shapes democracy, particularly how far the promise of monitory democracy can go.[29] This is the first of three essay prompts they have for the second half of the semester. The remaining two prompts are about surveillance and automation.

[27] *The Spy in Your Phone* (Al Jazeera World, 2021), https://www.aljazeera.com/program/al-jazeera-world/2021/1/6/the-spy-in-your-phone

; Nawal Al-Maghafi, *Weapons of Mass Surveillance*, BBC News, July 7, 2017, https://www.bbc.com/news/av/world-middle-east-40531967

[28] On malware, Fenton cites Ronald J. Deibert and Rafal Rohozinski, "Liberation vs. Control: The Future of Cyberspace," in *Liberation Technology: Social Media and the Struggle for Democracy*, ed. Larry Diamond and Marc F. Plattner (Baltimore, MD: Johns Hopkins University Press, 2012), 18–32. See Fenton, "The Internet of Me," 162.

[29] "Democracy" is of course a complicated term. As a historian of technology, I ask students to adhere to either a scholar's or historical actors' definition. For example, authors of these readings refer democracy to the processes in which members of a society can autonomously decide whom to vote for; individuals can express concerns freely over important issues in a society; or residents can exercise supervision over local or national governments.

Surveillance in Emerging Technologies

Surveillance, an influential feature of many social media technologies, is the theme of the second module of the second half of the semester. By now, students are getting more familiar with the rampant surveillance on social media and in other online worlds. They saw how a political campaign company could identify over 500 data points from each Facebook user. The nature of no-click hacking to locate and arrest dissidents through the spyware Pegasus delineated in *The Spy in Your Phone* often surprises students, even though they tend to associate it with their loved ones' experiences of becoming fraud victims after their mobile phones were compromised by other methods.

I then introduce more scholarly works and more cases about surveillance to develop their critical thinking of surveillance on social media. First, I assign Michel Foucault's writing about "Panopticism" and identify three features of this new type of surveillance that emerged in the late eighteenth century. Unseen and unverifiable gazes are essential in a panopticon, which offers a more economic and efficient way to exercise power by disciplining prisoners or other surveilled populations and automates the process of exercising power.[30] Students and I then work together to consider the following two questions in lectures and discussion sections: If Foucault were alive during the social media age, what types of new mechanisms in social media would he have identified? What types of theories and criticism would he have developed? Students often agree with me that the panoptic surveillance in the late eighteenth century is strikingly similar to how our personal data has been tracked by various pro-profit companies now. One difference that they tend to identify is that after the Cambridge Analytica-Facebook data scandal, some of us try not to be tracked by companies who want to have our data. Therefore, some of us do not discipline ourselves to comply with the entities who surveil us.

Right after "Panopticism," students watch the documentary *Coded Bias*, which elaborates Joy Buolamwini's widely cited research—some facial recognition systems have a lower accuracy rate when identifying women and darker-skinned faces than men and lighter-skinned faces.[31] This documentary supplies students with many cases about how surveil-

[30] Michel Foucault, "Panopticism," *Discipline and Punish: The Birth of the Prison* (New York: Vintage Books 1995), 195–228, translated by Alan Sheridan.

[31] *Coded Bias* (Shalini Kantayya, 2021), www.codedbias.com.

lance techniques are embedded with technology in our time. *Coded Bias*' director Shalini Kantayya not only exposes and analyzes cases that implemented certain facial recognition systems and AI tools and their violations of civil rights, but she also emphasizes advocacy to stop these violations. For example, on the documentary's website, toolkits for advocacy are available for educators and viewers to "demand transparency about government use of surveillance technology."[32] The film also features scenes in which Big Brother Watch, a non-profit organization directed by Silkie Carlo in the United Kingdom, helped pedestrians in London who were subject to police-implemented facial recognition systems. The example that my students found the most shocking is that a 14-year-old darker-skinned boy in school uniform was mistaken as a suspect by a facial recognition system implemented by plain clothes officers on the street. He was thoroughly searched and asked to give his fingerprints before the officers realized it was a mismatch. One of the many advocacy cases in the documentary focuses on residents of Atlantic Plaza Towers in Brownsville in Brooklyn, New York, who organized to stop their landlord from replacing key fobs with a facial recognition system and installing the system everywhere in the apartment building.

Coded Bias includes interviews with Joy Buolamwini, Cathy O'Neil, Zeynep Tufekci, Meredith Broussard, and other scholars and activists. Each of the scholars' monographs can be great readings that easily generate discussions and responses from students of this course. I also use their masterpieces for lecture preparation. Additionally, Kashmir Hill's 2024 *Your Face Belongs to Us: A Tale of AI, a Secretive Startup, and the End of Privacy* also offers an in-depth analysis of a facial recognition tech company, Clearview AI. It is one of several companies that continued to offer facial recognition technology to law enforcement when Google, Amazon, and Microsoft, after the arrest of George Floyd protesters in summer 2020, decided that "they won't sell or would at least pause police use of their facial recognition technology until there are federal laws on the matter."[33]

[32] See *Coded Bias*'s official website for a document titled "Activist toolkit," especially page 17, https://www.codedbias.com/takeaction

[33] Geoffrey A. Fowler, "Black Lives Matter could change facial recognition forever — if Big Tech doesn't stand in the way," *Washington Post*, June 12, 2020, https://www.washingtonpost.com/technology/2020/06/12/facial-recognition-ban/. See also CNBC, "The Fight Over Police Use of Facial Recognition Technology,"July 11, 2020, https://www.youtube.com/watch?v=oCwEYi_JjEQ

In addition to reviewing the advocacy emphasized in *Coded Bias*, I also introduce artworks created by Adam Harvey and others that critique the surveillance feature of facial recognition technology. His *HyperFace*, created in 2017, is a camouflage scarf that contains numerous human faces to overload a facial recognition system. In 2013, he also developed a hijab, *Stealth Wear*, that is made of silver-plated fabric to block thermal radiation and may protect the wearer from being seen in drone warfare.[34] In lectures, students seem to appreciate the creativity demonstrated by the camouflage scarf. I hope Harvey and others' artistic expression can inspire students to develop their own thoughts on surveillance in our time.

I then focus on an empirical study of algorithms, surveillance, and social inequality so that students learn more on how to approach surveillance issues with humanities and social sciences methods: Virginia Eubanks' *Automating Inequality: How High-Tech Tools Profile, Police, and Punish the Poor*, especially the case about a screening tool to evaluate the risk of child neglect and abuse cases in Pittsburgh.[35] Eubanks argues that the algorithms of the screening tool heavily rely on data available to the Allegheny County government, data of the residents with fewer resources who have been using or had used SNAP (the Supplemental Nutrition Assistance Program), public housing subsidies, Medicaid, and other similar public services. Therefore, the score generated by the screening tool about a poor family that has had a history of using public services would be higher than the score generated for an upper-middle-class family that has had little to no interaction with public service agencies. When a phone call reporting a suspicious case of child neglect or abuse in a residence is received, it is more likely that the screening tool would identify a poor family as a high-risk case—followed by an investigation—than it would identify a wealthy family as high risk. As many families in the county with fewer recources are African American and interracial families, the screening tool tends to identify families of color as high-risk cases. High-tech tools, as documented by Eubanks, were first implemented in poorer communities in the name of protection. Subsequently, these tools would be used for surveillance purposes. Historians of technology who are familiar with SCOT may find Eubanks' approach similar to the scholarship of

[34] See Adam Harvey's website at https://adam.harvey.studio/hyperface/ and https://adam.harvey.studio/stealth-wear/

[35] Virginia Eubanks, "The Allegheny Algorithm," *Automating Inequality: How High-tech Tools Profile, Police, and Punish the Poor* (New York: St. Martin's Press, 2018), 127-173.

social construction of knowledge, as Eubanks clearly demonstrates the flaws in the design of the algorithms in question, especially problematic variables chosen from data sets that focus on the poor.

By the end of this surveillance module, students have plenty of material to discuss how twenty-first-century cases can substantiate Michel Foucault's concept of panopticism, which was originally about a historical phenomenon in the late eighteenth century. This is the second essay prompt for the second half of the semester: students are asked to describe and analyze panopticon-like surveillance mechanisms in the digital world.

For contemporary issues about social media in relation to democracy and surveillance, I also discuss each case with the framework of social groups and their counternarratives. The dominant narratives about the "usefulness" of these new digital surveillance technologies are compelling, especially those constructed and supported by for-profit corporations. As opposed to for-profit corporations, the role of politicians should be guarding the public's interest and critically examining these dominant narratives. To me it is then important for undergraduate education instructors to help students identify social groups of various racial, gender, and socioeconomic background in these contemporary cases, and to explore whether they are crafting their counternarratives against dominant narratives. How do minorities and political dissidents narrate and respond to technologies with growing capacities to trace physical location, communications, and online-content? As in Eubanks' case, can individuals "talk back" to a score assigned by concealed algorithms? How do civil rights watch groups use their expertise to analyze the impact of loosely regulated emerging technologies on civil rights? How are those efforts different from cost and benefit analyses done by for-profit entities? Historians of technology not only tell stories as writers, but also as instructors, "be they progressivist or declensionist, triumphant or tragic."[36] For facial recognition systems especially, as legal regulations over this technology are evolving and uncertain, different groups attempt to shape the regulations, the market, and the technology's cultural connotations. Historians of technology, with our scholarly attention to the technical specifications of each artifact and their significance, are well-positioned to guide students to tease out these ongoing, murky entanglements of new digital technologies and society.

[36] Herzig, "*A Thing for Stories*," 617.

Cybernetics, Automation, and the History of AI

The third module for the second half of the semester is about algorithms, machine learning techniques,[37] and artificial intelligence that have recently automated many social decisions. In this module students are expected to identify authors' concerns over automation. At this moment, students can identify concerns and problems discussed in Eubanks' *Automating Inequality and* Shalini Kantayya's *Coded Bias.* For example, Eubanks emphasizes the automation power of algorithms to replace social workers to make decisions. She argues that the automation of evaluating child neglect and abuse risk still allows racial bias transferred into algorithms. Proponents of the screening tool deem the risk scores generated by computers and algorithms more objective than human reasoning, as they believe subjective judgment can be avoided if a decision is made by machines instead of humans. But in reality, human experiences of individual cases are undervalued in this automating process, and racial bias embedded in the data sets smoothly creeps into algorithms.

The new assigned readings for this module include Alexandrea J. Ravenelle's *Hustle and Gig* and the panel "Historical Perspective on Artificial Intelligence," organized by the American Historical Association (AHA) in November 2023, as one of its congressional briefing series.[38] Reading *Hustle and Gig*, students see that the sharing economy, or gig economy, including services provided by Uber, Airbnb, etc., is an extension of the movement to hire more temporary workers in corporations during the past decades. Corporations usually can forgo temporary work-

[37] On machine learning, Vladimir Cherkassky, a professor of electrical and computer engineering, has given two insightful lectures for my course over the past years. He takes a philosophy of science approach to understand the capabilities, limitations, and misconceptions in machine learning methods. See Vladimir Cherkassky, "Predictive Learning, Knowledge Discovery and Philosophy of Science," *Advances in Computational Intelligence: IEEE World Congress on Computational Intelligence, WCCI 2012, Brisbane, Australia, June 10–15, 2012. Plenary/Invited Lectures*, ed. Jing Liu, Cesare Alippi, Bernadette Bouchon-Meunier, Garrison W. Greenwood, Hussein A. Abbass (Springer-Verlag Berlin Heidelberg, 2012), 209–233.

[38] Alexandrea J. Ravenelle, *Hustle and Gig: Struggling and Surviving in the Sharing Economy* (Oakland, California: University of California Press, 2019). Matthew L. Jones, Jeffrey Yost, Janet Abbate, and Matthew Connelly, "Historical Perspective on Artificial Intelligence," American Historical Association's Congressional Briefings series, C-SPAN, October 20, 2023, https://www.c-span.org/video/?531280-1/historical-perspective-artificial-intelligence

ers' social security contributions, and health insurance, and other benefits. Therefore, it would be more economical for corporations to hire temporary workers when they can. Despite that tech companies' powerful algorithms can now automatically match service providers and customers. Ravenelle argues that these algorithms are "just the newest (technological) innovation[s] in treating workers shabbily."[39] In the past, I have also used two other academic works to discuss the elusive economic opportunities offered by the rise of the internet. "The Internet of Capital" chapter in *Misunderstanding the Internet* emphasizes the neoliberal nature of economic activities rendered by the internet.[40] Fred Turner's "Millenarian Tinkering: The Puritan Roots of the Maker Movement" unpacks the enthusiasm about makers in the context of economic uncertainties in today's United States.[41]

To illustrate long-lasting corporate interests in using digital technology for automation, I assign Jeffrey Yost's and Janet Abbate's presentations for the AHA's panel titled "Historical Perspective on Artificial Intelligence" in 2023.[42] In watching Yost's presentation, students can learn that journalists and scientists used terms such as giant brains and electronic brains to describe computers in the late 1940s and 1950s. They also learn that John McCarthy coined the term "artificial intelligence," as he and others needed to present their research in a way that would attract military and federal funding, including from the Advanced Research Projects Agency (ARPA), which initiated the building of the ARPANET, a subject of the earlier part of the semester. Through this history, Yost illustrates that "intelligence" is a metaphor that researchers evoked to articulate the utility of the discipline of artificial intelligence. Metaphors of intelligence reveal the hype of several waves of enthusiasm for artificial intelligence in past decades. Yost further stresses that artificial intelligence's application in automation and its alleged promise about labor-saving can be a better metaphor than "intelligence" for explaining why many corporations would like to embrace this technology.

Abbate's presentation also focuses on choices of terminology in narratives and counternarratives about artificial intelligence. She points out that

[39] Ravenelle, *Hustle and Gig*, 180.

[40] Des Freedman, "The Internet of Capital," *Misunderstanding the Internet*, 85–116.

[41] Fred Turner, "Millenarian Tinkering: The Puritan Roots of the Maker Movement," *Technology and Culture* 59, no. 4 supplement (2018): S160–S182.

[42] Jones, Yost, Abbate, and Connelly, "Historical Perspective on Artificial Intelligence."

the term artificial intelligence does not really describe what it does. If artificial intelligence systems have not demonstrated general-purpose intelligence, they could have been named after the specific aspects of intelligence they can display. Abbate also urges us to think about what counts as a success and what counts as a failure. For example, while the popularity (market share) or crowdsourced reviews can be an indicator to evaluate the success of an artificial intelligence system, it should be equally important to take into account whether the answers generated by an artificial intelligence system are correct or not. Abbate notes that in cases in which users request an artificial intelligence system to generate hate speech, the success of the user should be considered a failure of developers and their society. Both Yost and Abbate have reviewed and developed counternarratives about artificial intelligence, different from dominant narratives.

Kline's *Cybernetics Moment* especially helped me emphasize in the classroom the recent historical research on artificial intelligence offered by Yost, Abbate, and other historians of technology. In *Cybernetics Moment*, Kline has pointed out that early artificial intelligence researchers distinguished themselves from cyberneticians in the mid-1950s. As Paul Edwards has noted, it was the beginning of the split between the computer-brain metaphor (the concern of cybernetics) and computer-mind metaphor (the concern of artificial intelligence). Cyberneticians aimed to use mechanical devices to imitate how the human brain works, but for researchers who worked on what we now broadly call artificial intelligence, they meant to mechanize human reasoning. Thus, the research on cybernetics is not the predecessor of research on artificial intelligence. Their historical continuity is questionable.[43] However, Kline's emphasis on the analogy of humans and machines prompts me to stress in my teaching the power of metaphor and the obsession with automation in both history and in today's dominant narratives about artificial intelligence.[44] In his 2016 address Kline describes his research on cybernetics in relation to the theme of humans and machines:

[43] Kline, *Cybernetics Moment*, 157–165. For Paul Edwards' discussion of the split, I cited from Kline, *Cybernetics Moment*, 159.

[44] It also helped me conceptualize economists' crafting of an analogy of computer and economy. See Honghong Tinn, "Between 'Magnificent Machine' and 'Elusive Device': Wassily Leontief's Interindustry Input-output Analysis and its International Applicability," *Osiris* 38 (2023): 129–146.

> I tried to explain why so many scientists, engineers, and humanists in the U.S. in the 1950s and 1960s thought cybernetics could shed new light on the age-old question about the relationship between humans and machines. For them, the basis of cybernetics was a powerful analogy: that the principles of information-feedback machines, which explained how a thermostat controlled a household furnace, for example, could also explain how all living things—from the level of the cell to that of society—behaved as they interacted with their environment.[45]

Norbert Wiener, the protagonist of *Cybernetics Moment*, discussed automation in his *The Human Use of Human Beings* in 1950.[46] Kline has also noted Wiener's concerns over automation in his iterations of the undergraduate course "Inventing an Information Society" at Cornell University many years prior to his publication of *Cybernetics Moment* in 2015. In his 2016 address, he wrote:

> In 1950 Norbert Wiener wrote a popular book on the social implications of cybernetics that covers all these issues and is still in print today. In *The Human Use of Human Beings*, Wiener warned of the unemployment and other social ills that would result from the misapplication of cybernetics by authoritarian governments in regard to automatic factories and secrecy in military-funded research. Yet he also thought cybernetics would strengthen liberal democracy if it was applied to language, law, education, and government.[47]

Kline's *Cybernetics Moment*, Yost's and Abbate's analyses in the AHA panel discussion, and a wealth of recent research on the history of machine learning, pattern recognition, computer vision, and artificial intelligence are teaching resources for historians of technology that reveal how the idea of automation has been an integral part of the history of artificial intelligence.[48] The attempts to automate human decisions ranged from

[45] Kline, "Humans and Machines," 840.

[46] Norbert Wiener, *The Human Use of Human Beings: Cybernetics and Society* (Boston, MA: Houghton Mifflin, 1950).

[47] Kline, "Humans and Machines," 841.

[48] For example, see Stephanie Dick, "The Marxist in the Machine," *Osiris* 38 (2023): 61–81; Aaron Mendon-Plasek, 2023, "Irreducible Worlds of Inexhaustible Meaning: Early 1950s Machine Learning as Subjective Decision Making, Creative Imagining and Remedy for the Unforeseen," *BJHS Themes* 8: 65–80, https://doi.org/10.1017/bjt.2023.12; Aaron Gluck-Thaler, 2024. "Identity Instrumentalized: Pattern Recognition as an Epistemology of

those made by scientists of different disciplines to government agencies with various goals during the Cold War. These works might help students better understand the rise and fall of artificial intelligence in past decades and their respective political, cultural, economic, and technical contexts. Students can then compare the past to today.

Conclusion

In this chapter, I discuss several toolkits that can help us envision the roles of historians of technology in teaching emerging technologies in what we call "the information age." I first discuss the emphasis on "social groups" in teaching the history of the internet and other recent internet-powered technologies, a familiar framework for many historians of technology. The identification of the interpretations of technology by relevant social groups has encouraged students to engage with a close reading of each social group's narratives, and it has pushed students to evaluate multiple existing interpretations of a technology. In the context of today's social media and newer digital technologies, the category "social groups" helps students stay open-minded about possible directions a technology can evolve into.

Second, the framework of social groups also encourages students to consider "counternarratives," and to think beyond popular, dominant narratives about technologies, especially those provided by for-profit corporations. If students can recognize different types of narratives, they may begin to think with which narratives they are willing to side, and what roles they can play in the history of a technology. An anonymous student wrote to me before heading to a large tech company for an summer internship, noting that the course "opened my eyes to issues I had not considered before," and "I'll try to keep them in mind" during the internship.

Third, automation is an important theme through which students can make sense of dominant narratives about digital technologies, especially those developed by corporations seeking to replace human labor and human reasoning. Metaphor is an effective tool for us to analyze dominant narratives and counternarratives. In my teaching experience, undergraduate students are concerned about their career opportunities. Historians

Surveillance, 1960–66," *Technology and Culture* 65, no. 4: 1161–1193; Jeffrey R. Yost and Gerardo Con Díaz, "Epilogue: Artificial Intelligence—Braiding Irony, Paradox, and Possibility," in *Just Code: Power, Inequality and the Political Economy of IT*, ed. Jeffrey R. Yost and Gerardo Con Díaz (Baltimore, MD: Johns Hopkins University Press, 2025), 411-432.

can help students develop sophisticated analyses of how future jobs can be shaped by different factors, including the pro-profit goals of corporations, government regulations, and potentials and limitations of emerging technologies. Kline shared an anecdote with his undergraduate students when he taught "Inventing an Information Society," and I think the anecdote speaks to historians of technology's central role in undergraduate education. When he taught history of technology courses during the dot-com boom, some students hired by startup companies sat in the back row of the classroom and wrote computer programs in class. One day, he lectured about the 1960s tech industry boom and the subsequent crash and recession at the turn of the decade. The students who usually concentrated on the code they were writing raised their heads, looked at Kline, and puzzled over what he had just said. They might be wondering what a recession was, as it was an unimaginable concept during the dot-com boom. Even though historians of technology cannot predict the future, history of technology classes may equip students with concepts, ideas, and patterns that they might encounter in their careers.

In late March 2025, I showed my students the last five minutes of Carole Cadwalladr's 2019 TED talk. I have showed this short video clip in class every year since 2021. As *The Guardian*'s journalist, Cadwalladr was one of the most influential investigative reporters to reveal the Cambridge Analytica-Facebook data scandal. In the talk, she advocated for more attention to large tech companies' data privacy ignorance, which is exercising an adverse power over democracy. In her closing remarks, she warned, "And this is not a drill – it's a point of inflection. Democracy is not guaranteed, and it is not inevitable, and we have to fight and we have to win and we cannot let these tech companies have this unchecked power. It's up to us – you, me and all of us. We are the ones who have to take back control."[49]

A student applauded after Cadwalladr's closing remarks in late March 2025, as if he was one of the emotionally touched TED talk audience in

[49] Carole Cadwalladr, "Facebook's Role in Brexit — and the Threat to Democracy," TED talks, April 2019, https://www.ted.com/talks/carole_cadwalladr_facebook_s_role_in_brexit_and_the_threat_to_democracy

the video. Perhaps more students would have wanted to applaud to express their agreement with Cadwalladr, but keep in mind that applauding is not a regular thing students do during a general education class in a huge lecture hall. I should have joined the student to applaud, too. But I tend to refrain from improvising too much when lecturing to a large class. Instead, I thanked the student for his participation, and I shared with the class that I must have also clapped my hands when I watched the video earlier that day, prepping the lecture in my office. I appreciate the student' applauding as it is encouraging to know they like and take away something from the course materials. As Herzig and Kline have noted, storytelling by historians of technology "engage[s] the morality, the politics, of technology, whether we want them to or not."[50] Historians of technology can contribute to actions "to take back control" of democracy, as Cadwalladr termed. By teaching these important works and methods, we can empower students to assess emerging technologies in a deliberative and responsible way.

Bibliography

Videos and Websites

Harvey, Adam. Personal website at https://adam.harvey.studio/hyperface/ and https://adam.harvey.studio/stealth-wear/.

Al Jazeera World. 2021. *The Spy in Your Phone*. January 6. https://www.aljazeera.com/program/al-jazeera-world/2021/1/6/the-spy-in-your-phone.

Al-Maghafi, Nawal. 2017. *Weapons of Mass Surveillance*. BBC News, July 7. https://www.bbc.com/news/av/world-middle-east-40531967.

Cadwalladr, Carole. "Facebook's Role in Brexit — and the Threat to Democracy," TED Talks, April 2019. https://www.ted.com/talks/carole_cadwalladr_facebook_s_role_in_brexit_and_the_threat_to_democracy.

CNBC. 2020. "The Fight Over Police Use of Facial Recognition Technology." July 11. https://www.youtube.com/watch?v=oCwEYi_JjEQ.

Earp, Jeremy. 2018. *Digital Disconnect: How Capitalism is Turning the Internet against Democracy* [film], featuring Robert W. McChesney.

Jones, Matthew L., Jeffrey Yost, Janet Abbate, and Matthew Connelly. 2023. "Historical Perspective on Artificial Intelligence." American Historical Association's Congressional Briefings series. C-SPAN, October 20. https://

[50] Kline, "Foundational Stories," 119.

www.c-span.org/video/?531280-1/historical-perspective-artificial-intelligence.

Kantayya, Shalini. 2021. *Coded Bias.* www.codedbias.com.

Books and Articles

Abbate, Janet. 2000. *Inventing the Internet.* Cambridge, MA: MIT Press.

Adamson, Greg, Ronald R. Kline, Katina Michael, and M. G. Michael. 2015. "Wiener's Cybernetics Legacy and the Growing Need for the Interdisciplinary Approach [Scanning Our Past]." *Proceedings of the IEEE* 103, no. 11: 2208–2214.

Cherkassky, Vladimir. 2012. "Predictive Learning, Knowledge Discovery and Philosophy of Science." In *Advances in Computational Intelligence: IEEE World Congress on Computational Intelligence, WCCI 2012, Brisbane, Australia, June 10–15, 2012. Plenary/Invited Lectures.* Edited by Jing Liu, Cesare Alippi, Bernadette Bouchon-Meunier, Garrison W. Greenwood, and Hussein A. Abbass, 209–233. New York: Springer-Verlag Berlin Heidelberg.

Deibert, Ronald J. and Rohozinski, Rafal. 2012. "Liberation vs. Control: The Future of Cyberspace." In *Liberation Technology: Social Media and the Struggle for Democracy.* Edited by Larry Diamond and Marc F. Plattner, 18–32. Baltimore, MD: Johns Hopkins University Press.

Dick, Stephanie. 2023. "The Marxist in the Machine." *Osiris* 38: 61–81.

Eubanks, Virginia. 2018. "The Allegheny Algorithm." *Automating Inequality: How High-tech Tools Profile, Police, and Punish the Poor.* New York: St. Martin's Press, 127-173.

Fenton, Natalie. 2016. "The Internet of Me (and My 'Friends')." *Misunderstanding the Internet.* Co-authored by Curran, James, Natalie Fenton, Des Freedman, 145–172. London: Routledge, 2nd Edition.

Forster, E.M. "The Machine Stops," 1909. https://www.gutenberg.org/cache/epub/72890/pg72890.txt

Foucault, Michel. 1995. "Panopticism." *Discipline and Punish: The Birth of the Prison.* Translated by Alan Sheridan. New York: Vintage Books, 195–228.

Fowler, Geoffrey A. 2020. "Black Lives Matter could change facial recognition forever — if Big Tech doesn't stand in the way." *Washington Post*, June 12. https://www.washingtonpost.com/technology/2020/06/12/facial-recognition-ban/.

Freedman, Des. 2016. "The Internet of Capital." *Misunderstanding the Internet.* Co-authored by Curran, James, Natalie Fenton, Des Freedman, 85–116. London: Routledge, 2nd Edition.

Gluck-Thaler, Aaron, 2024. "Identity Instrumentalized: Pattern Recognition as an Epistemology of Surveillance, 1960–66." *Technology and Culture* 65, no. 4: 1161–1193.

Herzig, Rebecca. 2009. "A Thing for Stories," *Technology and Culture* 50, no. 3: 613–622.

Kaiser, Brittany. 2019. *Targeted: The Cambridge Analytica Whistleblower's Inside Story of How Big Data, Trump, and Facebook Broke Democracy and How It Can Happen Again*. New York: HarperCollins.

Keane, John. 2009. "Monitory Democracy and Media-saturated Societies." *Griffith Review* 24: 47–69.

Kline, Ronald R. 2017. "Da Vinci Medal Address: Humans and Machines." *Technology and Culture* 58, no. 3 (2017): 835–845.

Kline, Ronald R. 2000. *Consumers in the Country: Technology and Social Change in Rural America*. Baltimore, MD: Johns Hopkins University Press.

Kline, Ronald R. 2006. "Cybernetics, Management Science, and Technology Policy: The Emergence of 'Information Technology' as a Keyword, 1948–1985." *Technology and Culture* vol. 47 no. 3: 513–535.

Kline, Ronald R. 2011. "Cybernetic as a Useable Past, Review of Andrew Pickering: *The Cybernetic Brain: Sketches of Another Future*." *Metascience* 20: 519–524.

Kline, Ronald R. 2013. "Presidential Address: Foundational Stories." *Technology and Culture* 54, no. 1: 117–129.

Kline, Ronald R. 2015. *The Cybernetics Moment: Or Why We Call Our Age the Information Age*. Baltimore, MD: Johns Hopkins University Press.

Kline, Ronald R. and Trevor J. Pinch. 1996. "Users as Agents of Technological Change: The Social Construction of the Automobile in the Rural United States." *Technology and Culture* 37, no. 4: 763–795.

Kranzberg, Melvin. 1986. "Technology and History: 'Kranzberg's Laws.'" *TechnologyandCulture*27,no.3:544–560.https://doi.org/10.2307/3105385.

McChesney, Robert W. 2013. *Digital Disconnect: How Capitalism is Turning the Internet against Democracy*. New York: The New Press.

Medina, Eden. 2006. "Designing Freedom, Regulating a Nation: Socialist Cybernetics in Allende's Chile." *Journal of Latin American Studies* 38, no. 3: 571–606.

Medina, Eden. 2011. *Cybernetic Revolutionaries: Technology and Politics in Allende's Chile*. Cambridge, MA: MIT Press.

Mendon-Plasek, Aaron. 2023. "Irreducible Worlds of Inexhaustible Meaning: Early 1950s Machine Learning as Subjective Decision Making, Creative Imagining and Remedy for the Unforeseen." *BJHS Themes* 8: 65–80. https://doi.org/10.1017/bjt.2023.12

Nye, David E. 2003. *America as Second Creation*. Cambridge, MA: MIT Press.

Pinch, Trevor J. and Wiebe Bijker. 1987 [1984]. "The Social Construction of Facts and Artifacts: Or, How the Sociology of Science and the Sociology of Technology Might Benefit Each Other." In *The Social Construction of Technological Systems*. Edited by Wiebe Bijker, Thomas P. Hughes, and Trevor J. Pinch, 17–50. Cambridge, MA: MIT Press.

Ravenelle, Alexandrea J. 2019. *Hustle and Gig: Struggling and Surviving in the Sharing Economy*. Oakland, California: University of California Press.

Tinn, Honghong. 2023. "Between 'Magnificent Machine' and 'Elusive Device': Wassily Leontief's Interindustry Input-output Analysis and its International Applicability." *Osiris* 38: 129–146.

Townsend, Robert B. 2021. "Has the Decline in History Majors Hit Bottom?" *Perspectives on History*, February 23. https://www.historians.org/perspectives-article/has-the-decline-in-history-majors-hit-bottom-data-from-2018-19-show-lowest-number-since-1980-march-2021/

Turner, Fred. 2005. "Where the Counterculture Met the New Economy: The WELL and the Origins of Virtual Community." *Technology and Culture* 46, no. 3: 485–512.

Turner, Fred. 2006. *From Counterculture to Cyberculture: Stewart Brand, the Whole Earth Network, and the Rise of Digital Utopianism*. Chicago: University of Chicago Press.

Turner, Fred. 2018. "Millenarian Tinkering: The Puritan Roots of the Maker Movement." *Technology and Culture* 59, no. 4, supplement: S160–S182.

Wiener, Norbert. 1950. *The Human Use of Human Beings: Cybernetics and Society*. Boston, MA: Houghton Mifflin.

Yost, Jeffrey R. and Gerardo Con Díaz, 2025, "Epilogue: Artificial Intelligence—Braiding Irony, Paradox, and Possibility." In *Just Code: Power, Inequality and the Political Economy of IT*. Edited by Jeffrey R. Yost and Gerardo Con Díaz, 411–432. Baltimore, MD: Johns Hopkins University Press.

CHAPTER 9

Science Outside the Lab: Ethics Training Through Science Policy Immersion

Jameson M. Wetmore

Introduction

During the late 1990s, I worked closely with Ron Kline to find ways to use basic lessons from STS to inform engineering ethics.[1] We were inspired by Technology Studies and adjacent scholars who had begun to illustrate the

[1] William Lynch, Ronald Kline, and Jameson M. Wetmore, "Book reviews of 'Engineering Ethics: Balancing Cost, Schedule, and Risk-Lessons Learned from the Space Shuttle' and 'The Challenger Launch Decision,'" *ISIS* 89, no. 4 (1998): 761–763; Ronald R. Kline, "Using history and sociology to teach engineering ethics," *IEEE Technology and Society Magazine* 20, no. 4 (2001); William T. Lynch and Ronald Kline, "Engineering Practice and Engineering Ethics," *Science, Technology, & Human Values* 25, no. 2 (2000).

J. M. Wetmore (✉)
School for the Future of Innovation in Society, Arizona State University, Tempe, AZ, USA

Center for Engagement and Training in Science & Society, Arizona State University, Tempe, AZ, USA
e-mail: Jameson.wetmore@asu.edu

J. L. Lieberman, A. Mohsin (eds.), *Technological Humanities*,
https://doi.org/10.1007/978-3-032-24741-4_9

complex nature of engineering practice.[2] Thanks to insights from the Sociology of Scientific Knowledge we knew that defining facts—including engineering facts—was not the simple process presented in many engineering textbooks.[3] And we knew from Ron's own work, and of course that of Trevor Pinch and others, that it was groups of people who shaped technology.[4] Ron took his joint appointment in Science & Technology Studies and Electrical Engineering very seriously. In his work to bridge the two disciplines he had come to the conclusion that engineers could better understand the complexity in which they worked and the impact that their decisions had if they had some STS tools in their toolbox.

The spirit and goal of that work has never left me. I have continued to try to find ways to take what STS scholars have learned about engineering practice and the impacts of technologies to help engineers make more informed decisions.[5] And while my early training was focused on engineering, over the years I have widened my purview a bit. I have found that when I walk through a lab it is often difficult to tell who is officially a scientist and who has an engineering degree. I have found that many scientists engage in issues that are analogous to those covered in engineering

[2] Donald MacKenzie, *Inventing accuracy: A historical sociology of nuclear missile guidance* (MIT Press, 1990).

[3] Trevor J. Pinch and Wiebe E. Bijker, "The social construction of facts and artefacts: Or how the sociology of science and the sociology of technology might benefit each other," *Social Studies of Science* 14, no. 3 (1984); Charles E. Harris, Michael S. Pritchard, and Michael Rabins, *Engineering Ethics: Concepts and Cases, 1st edition* (Wadsworth Publishing, 1994).

[4] Ronald Kline and Trevor Pinch, "Users as agents of technological change: The social construction of the automobile in the rural United States," *Technology and Culture* 37, no. 4 (1996); Wiebe E. Bijker, Thomas P. Hughes, and Trevor J. Pinch, eds, *The Social Construction of Technological Systems: New Directions in the Sociology and History of Technology* (MIT Press, 1987).

[5] Deborah G. Johnson and Jameson M. Wetmore, "STS and Ethics: Implications for Engineering Ethics," in *New Handbook of Science and Technology Studies*, ed. Ed Hackett, Olga Amsterdamska, Michael Lynch, and Judy Wajcman (MIT Press, 2007); Deborah G. Johnson and Jameson M. Wetmore, *Technology & Society: Building our Sociotechnical Future, Second Edition* (MIT Press, 2001);

Jameson M. Wetmore, Nalini Chhetri, and Mark Henderson, "Engineering in unfamiliar places: Two case studies for those who want to change the world," *IEEE Technology and Society* 44, no. 1 (2025); Thomas S. Woodson, Matthew Harsh, Michael J. Bernstein, Susan Cozzens, Jameson Wetmore, and Rafael Castillo, "Teaching community engagement to engineers via a workshop approach," *Journal of Professional Issues in Engineering Education and Practice* 145, no. 4 (2019).

ethics and the impacts that scientists can have on the world can be as profound as those of engineers.

In this chapter I will describe a program I co-developed that has proven to be a powerful way to help scientists and engineers not only begin to understand many of the basic insights enabled by STS, but to find ways to put them into practice. It is far from a traditional STS classroom experience. For instance, the program does not include philosophical theories or instruction in what Joe Herkert and Ron have called micro- and macroethics.[6] Nor do the students read Langdon Winner or articles about bicycles. But they learn to appreciate the complexities that STS scholars have so beautifully illuminated that are usually hidden from scientists and engineers in their university training.

My colleague, Ira Bennett, and I gave the program the title "Science Outside the Lab" because our target audience largely understood how science and engineering work inside the lab and we wanted to help them begin to understand how science and engineering is used by, engaged with, and influences the wide variety of actors outside the laboratory. Our goal was to give budding young scientists and engineers the tools they needed to increase the chances that their work ultimately makes the world a better place.

What we found was that in order to help PhD students understand "science outside the lab" it was incredibly helpful to get them physically outside of their labs. Getting them away from their day-to-day experiments was key to helping them be open to new complex ideas. Engaging with practitioners who can offer personal examples of how they bridge the social, ethical, and technical aspects of the world has proven more effective than theoretical case studies. And putting them in spaces where engineering and science are engaged with in a very different way went a long way to helping them see and appreciate the complexities of our socio-technical-political world.

[6] Joseph R. Herkert, "Ways of thinking about and teaching ethical problem solving: Microethics and macroethics in engineering," *Science and Engineering Ethics* 11, no. 3 (2005); Ronald R. Kline, "Engineering Case Studies: Bridging Micro and Macro Ethics," *IEEE Technology and Society Magazine* 29, no. 4 (2010); Heather E. Canary, Joseph R. Herkert, Karin Ellison, and Jameson M. Wetmore. "Microethics and macroethics in graduate education for scientists and engineers: Developing and assessing instructional models," *2012 ASEE Annual Conference & Exposition* (2012).

The Science Outside the Lab Approach

The Science Outside the Lab (SOtL) program is, on the face of it, an immersive dive into US Federal Science and Technology Policy.[7] We recruit cohorts of 15 participants who then spend one to two weeks in Washington, DC meeting with policymakers to learn what they do, how they got there, and the basics of how science policy works. In any given program we might meet with regulators, funders, lobbyists, journalists, museum curators, patent lawyers, federal judges, National Academies program managers, and congressional staffers. We've had a diversity of participants over the years, but the standard version of the program is focused on science and engineering students who are working on their doctorate.

While the official focus of the program is on science and technology policy—including the ways in which science and engineering are used to inform policy and the ways in which policy shapes, supports, and directs science and engineering—the program offers lessons that are much more broadly applicable. In many ways the program is a crash course in how to understand the relationships between science, technology, and society. And the program is more than just an introduction to these ideas but also shows participants how to integrate that knowledge into practice and make real change happen.

What I have found through previous research and confirmed through this program is that good S&T policymakers know a fair amount of STS. They may not be trained in STS, but they are in many ways STS practitioners.[8] They live and work at the border of scientific theory and daily practice. And as they become skilled at their jobs they develop an understanding of basic STS findings like how technology and society are interwoven, the ways in which technical decisions are bound up with values, and how different people can look at the same thing and see something different. If they came to their jobs thinking that technology and society were strongly delineated and clearly separated, their day-to-day experience had long

[7] We frequently abbreviate the sometimes cumbersome "Science Outside the Lab" with the acronym "SOtL." Typically we pronounce it as "sow-towl," which is also the way you pronounce "sotol," a beautiful plant that is native to the Chihuahuan Desert, but also grows well in Tempe, Arizona, the home of Arizona State University.

[8] Jameson M. Wetmore, "Engineering with uncertainty: Monitoring air bag performance," *Science and Engineering Ethics* 14, no. 2 (2008); Jameson M. Wetmore, "Delegating to the Automobile: Experimenting with Automotive Restraints in the 1970s," *Technology and Culture* 56, no. 2 (2015).

since disabused them of those myths. In many ways, the best science and technology policy practitioners have trained themselves in basic STS ideas by reflecting on their work at the interface of science, engineering, and political decision-making.

In addition to their knowledge of how to put ideas into action, science policy professionals are especially good vectors for helping young scientists and engineers better understand the complex sociotechnical world we live in because they share much of the same training. Most of the science and technology policy professionals in Washington, DC have a PhD in a science or engineering field and many spent significant time in labs very similar to those our participants normally inhabit.[9] This common background means that they not only know how to bridge the gap between where the students are and a broader view of the world, they are also more likely to be trusted when they describe science behaving in ways that are contrary to the myths commonly propagated to graduate students.

In what may seem to some an odd pedagogic decision, we have no specific learning objectives for the program. Our attitude is that there are so many things these students could learn that they will learn significantly more if they are allowed to explore what they want to explore. Thus, as much as those who run the program are instructors, they are also facilitators whose job it is to get these two groups together and create the space for them to learn from each other. That approach has led us to two of the basic rules of the program: there are no lectures and there are no PowerPoint presentations. Each speaker is given up to 15 minutes to introduce themselves, where they came from, how they got there, and what they do. At that point we open it up to questions from the participants.

At the outset of each program, we provide students some basic instruction in how to ask questions. We explain that many DC policymakers are often asked gotcha questions and have learned to defend themselves from outside attacks. This means that they are used to being asked hard questions and thus participants shouldn't shy away from diving into topics they are genuinely interested in. We have found that when our participants are truly motivated by intellectual and ask their questions politely and succinctly, they often get very interesting answers. Sometimes our speakers decline to answer or expertly avoid particularly sensitive areas. But even those instances provide a very important learning experience. This

[9] James Austin, "Paths to Science Policy," *Science.org*. February 28, 2003.

approach significantly increases the chances that we will learn more than can be found on the websites of the organizations we visit.

Over and over again we have found that PhD students have an incredibly powerful intellectual curiosity that extends far beyond whatever specific niche of academia they usually work on. When we give them a chance to ask incredibly knowledgeable individuals questions they've always wanted answered and never thought they'd get a chance to ask, the result more often than not is an incredibly interesting conversation.

Where we have these conversations is also very important. We usually have a "home base" in Washington, DC where we start and finish our week. Recently we've had the luxury of a reserved classroom at Arizona State University's Ambassador Barbara Barrett & Justice O'Connor Washington Center, which is centrally located just a couple blocks northwest of the White House. The space gives us a chance to brief students and prepare them for the week. We can also use the room to host speakers.

Whenever possible, however, we travel to the offices of the various speakers we meet with. We believe it is important for the students to not just grapple with powerful ideas, but to experience the contexts in which they are advanced and play out. This helps students experience the geographies of power in Washington, DC.

The benefits of moving around the District can be illustrated in the story of one especially memorable day in the program several years ago. In the early afternoon we were hosted by a well-funded lobbying firm. We were greeted by an impeccably dressed woman who escorted us to a beautiful board room and offered coffee, tea, and soda. Our speaker was as smooth as silk, made us all feel very important, and went a long way to convincing us that his organization was a force for good in the world as we looked out at a beautiful view of the Potomac. We then went directly to a small environmental organization. Their board room was not only tiny, it had no windows and also served as the organization's library. Our hosts were gracious and kindly let us know that if we needed the bathroom they could loan us a key and it was only a floor away. They had clearly poured their heart and soul into the job because it was a cause they truly believed in. The contrast between these two scenes was experienced firsthand by the students and went a long way toward helping them understand how the city works.

The Science Outside the Lab Participants

When I first ran this program in 2007 a number of our students came along either because their PI encouraged them to or they were looking for a break from the monotony of lab work. I sometimes describe the stereotypical participant as someone who goes to Washington thinking that if there were just a few policymakers with PhDs in the sciences, then most of the world's problems would be solved. When they get there, they meet with dozens of people with science and engineering PhDs, develop an enormous respect for them, and begin to understand how complicated it is to juggle multiple values and make decisions that are both widely supported and effective.

Many of the students in those early cohorts underwent transformational changes. Their eyes were opened to things they never expected to see. We had students who were seriously jarred by the idea that the pursuit of science did not automatically lead to public good. They had always been taught that science was inherently good and would always benefit the world. To realize that was not necessarily true caused significant emotional stress for a number of students. We found it incredibly important to help them through it and support them both intellectually and emotionally during and after the program ended. Because these learning experiences were so dramatic and caused the students to reevaluate just about everything they thought they knew, it was an incredibly gratifying program to coordinate.

As the program has matured, so have the students. Over at least the past five years I have witnessed a significant change in the students who apply for and participate in the program. Many of them have long since rejected the idea that science, technology, and values are separate endeavors. Their eyes have already been opened to at least some of the world's complexity and they can feel frustrated and confined by the training they have received. They know that the world is more complicated than has been presented to them in classes and even their lab experience. They are struggling to understand and operationalize very complex ethical considerations that they have little to no training to handle. And they are looking for ways to integrate consideration of outcomes, social justice, and public good into their work. I'm incredibly heartened to say that graduate student desires to link scientific and engineering work with broader social goals and issues of justice is stronger than ever. Many of them apply each year to the Science Outside the Lab program as one way to answer this call.

The impact of the program has not gone unnoticed by many of these students' faculty advisors. In the early days there were many PIs who felt that the program was an unnecessary distraction from the work the students should be doing in their labs. There were times when I had to lean on senior faculty who knew the program well to convince other faculty to let their students participate.

When students report back their personal experiences in the program, however, most PIs see its value. One time at an on-campus restaurant an alum introduced me to their advisor. The advisor quickly thanked me. He had learned that we had spent half a day at the National Science Foundation learning how and why its program managers make the decisions they do about which grants to fund. The advisor believed this was invaluable experience and said I was welcome to take any of his students any time I wanted.

Ultimately it was discussions like these between PIs and their returning students that led to conversations about whether a program could be put together for faculty. After a few years of preparations, we led our first faculty focused version in May 2023 and a second one in May 2025. The fundamental difference between the two programs is that in order to better fit in faculty schedules we reduced their program to three days. To help focus their programs we started with the assumption that they wouldn't have gotten as far as they already had in academia without a basic understanding of government funding, so we focused more on ways to better integrate their scientific findings into the policymaking process. The programs we've run with faculty have been quite successful and, in many ways, we've found that faculty have almost as much to learn as their students. For the most part in this chapter, however, I'll focus on the graduate student experience.

What Do They Learn?

I did state earlier that we don't have any specific learning outcomes in mind when we create the program. But we do have some broad learning goals and through the speakers we choose we are often able to ensure that we cover a number of key areas in both science and technology policy and STS.

Some of these are at the procedural level and help students understand the nuts and bolts of science policy. For instance, we want students to learn the basic process that federal agencies use to decide which proposals to fund and which ones not to. We want students to appreciate that regulators can't regulate everything so they have to find techniques to maximize their impact. We introduce students to people with a wide variety of perspectives to help them recognize that supporting scientific R&D has, at least prior to 2025, traditionally been a bipartisan effort. We show them that people with PhDs in science and engineering have a huge amount of input into government policies and work to make sure that scientific research informs policy decisions. And we meet with museum curators and journalists to help students recognize the importance of engaging with the public at various steps in the process of science.

But in many ways these general lessons don't quite get at the heart of what I hope the students get out of Science Outside the Lab. The stories our speakers share demonstrate time and again the basic STS lesson that values don't just matter, they're integral to any discussion or understanding of science and technology policy. The stories we hear illustrate a wide variety of ways in which values shape policy decisions and how those decisions in turn shape values.

In one iteration of SOtL we reenacted a congressional hearing. We had students develop five-minute pitches for research projects they thought would be worth US government investment. Then we were lucky enough to have congressional staffers volunteer to roleplay as members of Congress. In the House Science Committee Hearing room our students stood at the podium and offered their pitches. The "Members" would then ask the types of questions that are often asked in that very same room. In one memorable exchange a graduate student pitched an idea about funding a specific area of biochemistry. One of the Members nodded and smiled and then asked a simple and direct question: "I represent a rural district of the Great State of Nebraska. Tell me how a cattle rancher in western Nebraska will benefit from your proposal." The graduate student stood there stunned. They realized they had no answer to that question. As the stun wore off they, and their colleagues, subsequently realized that while the question had initially seemed absurd to them, in order to justify Congressional funding, in order to justify their idea as a legitimate public good, they did actually need a good answer to that question.

In our more recent programs we have worked on this specific issue through a personal professional development activity. At the beginning of the first week, in order to prepare for our many meetings and get the students to begin to look at the world in a new way, we have them practice introducing themselves. We use a simple technique of having participants explain their work to each other in pairs. Initially we give them two minutes to introduce themselves and their work. Their partner gives them feedback and then we decrease the time they have to present. After a few iterations, each student eventually develops a 20 second introduction.

Over the course of our program we might meet 50 policymakers, all with a strong background in science or engineering. But it's unlikely that any of them will be experts in the specific subfields that our students spend most of their time thinking about. As such students must develop introductions that are an important first step in connecting with a learned but non-expert audience. Many people think that the major gulf of communication between scientists and the general public is the jargon scientists use. And while recognizing jargon and eliminating it is an important step, over the years I've come to believe that it is the difference in values that is the bigger divide. In order to give a powerful introduction, the students learn that it is not enough to explain what they do, they need to make the link for their listener to help them understand why they should care.

Ultimately the twenty second introductions our students develop include: (1) the general areas they work in (e.g. material science or electrical engineering); (2) the mechanisms they focus on (e.g. carbon nanotubes or lithium batteries); and most crucially (3) the solutions to broadly understood problems that their work could address (e.g. more fuel-efficient airplanes or electric cars with greater range). The careful refinement of very short introductions offers the practicality of allowing a group of 15 students to introduce themselves in as little as five minutes. This quickly gives our speakers a general sense of where our students are coming from and time to ask more probing questions if there are specific things they are interested in exploring further. For our students, it is often their first foray into linking their work with values in an effort to explain why they matter to a broad audience. It is an important lesson that shows the benefit of being able to both see and clearly articulate those links.

One of the more difficult lessons the students usually learn is that science and engineering inform policy decisions, but rarely ever decide the specific outcomes of policy discussions. As with the "Member from Nebraska" example above, students learn over and over again that different people bring very different values to policy discussions and many of those values are completely valid. Students learn to respect that what constitutes a social good cannot be determined specifically through scientific inquiry. What counts as a social good can only be arrived at through a negotiation among different people with different values. There is not a linear path from "scientific discovery" to "social good."

While some of the participants start the program with the hope that more science will solve the world's problems, they begin to understand that scientific findings can fuel discussions, but cannot determine the outcome of discussions. Interestingly many students develop a healthy respect for this process. They quickly grasp that it is both incredibly complicated and important and they come to significantly appreciate the people who have the ability to simultaneously juggle values and science as they try to find solutions.

Ultimately the Science Outside the Lab program does not give a complete overview of engineering ethics. It doesn't go into the details of micro-ethics of why they shouldn't lie, cheat, or steal. And it doesn't help engineers know what to do when their boss asks them to "fudge it just a little." But it does help students understand the context and complexity within which science and engineering are done and the impacts they can have. It helps them to see things they simply couldn't see before. And it helps them to envision how they could play a direct role in "moving the needle." For students with a genuine desire to use their technical skills to make the world a better place, these lessons are invaluable. SOtL introduces students to STS concepts in a concrete and practical way that helps them engage with the relationships between science, technology, and values.

Impacts

We haven't been able to exactly pinpoint all the ways in which students are transformed by their experience with SOtL. But we do have a wide variety of evidence that the impact is powerful. We can see it through firsthand experience, pre- and post-testing, student surveys, and the career trajectories of our graduates.

Occasionally I get a "teacher's high" in a particular class or program I run. It's that feeling we instructors get when we recognize that students are opening their eyes and seeing the world in a new and enlightened way for the first time. One reason I am drawn back to running this program again and again is because it never fails to give me that teacher's high. A week is a relatively short time, but for some students it is clearly transformational. They start the program not quite knowing why they are there or where they are going and by the end they are asking piercing and important questions of our speakers to really unpack the issues underlying policy decisions.

Through post program discussions and surveys, many of our students are clearly able to articulate how they have changed. For instance, a student in our 2023 cohort explained their maturation through the program in the following way:

> [T]here are times as scientist doing research we feel like we know everything. I definitely thought with this program that I'd learn pathways to learn how to best promote my research for funding opportunities. But throughout the processes and people we met there were times when it was clear that we as the direct scientist maybe aren't the best vehicles for advocating for scientific policy change, especially if there are already people doing the work and setting the stage going through more official channels. Working with those people or giving the tools to aid in their advocation of change could be the better way than advocating directly sometimes.

This student discovered a productive humility through the program—recognizing that while science and technology are important, translating it to the policy world is crucial as well. This is a key lesson we hope our students get from the program and a key lesson from STS in general.

In line with many other engineering ethics pedagogies, our hope is that the students in our programs not only better understand complexity and the various value issues in it, but also transform their approach to the world to better promote social good. Thanks to assistance from Mary Aleta White of ASU's College Research Evaluation Services Team, we conducted a pre/post survey for the students in our 2019 program. We were very surprised to find that prior to the program only 20% agreed or strongly agreed with the statement: "The knowledge I provide should be used to help solve societal challenges." We were relieved to find that in the post survey, there was 100% agreement with the statement.

A student in our 2019 program reflected back on their experience and described not just what they learned, but how it changed their understanding of their responsibilities:

> I view my roles as a scientist and engineer as two-fold. First, it is my responsibility to conduct research and develop technologies that benefit society or satisfy a reasonable unmet need. Secondly, it is important that I communicate my work, not only to my peers, but also to policymakers and the public at large. Additionally, I should take on an active role in addressing policy issues pertaining to my scientific field as well as those issues in which I can provide an educated opinion. I should do so by writing in periodicals and

publications, as well as by communicating directly with stakeholders and decision makers.

Especially heartening for me is the fact that the student was able to clearly link their new understanding of responsibility with actionable and practical ways on how to carry out that responsibility.

Several years ago, we developed a few mechanisms to try to produce a quantitative measure of this learning process.[10] One approach we took was to have students draw out concept maps before and after the program. In the concept map students would draw what they saw as the "relationship of science and engineering to the government." At the most basic level we found that the number of nodes and links that students drew increased by 60% from their pre-program map to their post-program map. We interpreted this as evidence that students were much better able to appreciate the socially constructed nature of science policy and had moved well beyond a linear understanding of the impacts of science and engineering.

We also find that we can measure the impact on students by the ways in which their careers are inflected by the program. Most of our early students participated in the trip largely out of curiosity. Many of those students moved onto successful careers in academia and industry. But by the mid-2010s the program was sought out by students who were interested in a career in science policy and were looking for a solid next step in that direction. They knew that they needed training in how to navigate the complex world they could sort of see, but did not yet understand. We ultimately have been quite successful in being a spring board for those students.

For instance, of the 15 students who participated in the 2017 program, 3 went on to the prestigious AAAS Science & Technology Policy Fellowship, 2 went on to be Mirzayan Fellows at the National Academies of Science, and 2 went on to S&T Policy Fellowships for state governments. As of my writing this one of them is a Senate staffer, one is an advisor at the US Department of State, two of them are contractors for the Department of Energy, and one is a patent agent. Through SOtL we have helped young scientists transform into policy professionals that can span multiple disciplines, ideas, and values.

[10] Michael J. Bernstein, Kiera Reifschneider, Ira Bennett, and Jameson M. Wetmore. "Science outside the lab: helping graduate students in science and engineering understand the complexities of science policy," *Science and Engineering Ethics* 23, no. 3 (2017).

For the record, since 2021 we've also had three students who were pursuing PhDs in the social sciences and STS related fields when they participated in Science Outside the Lab. One of them has been publishing in policy journals as an assistant professor, one is now doing technology policy work at the United Nations, and one is an AAAS S&T Policy Fellow at the State Department. While this is a small sample size, it appears that the program can also help STS scholars make the transition to policy relevant work as well.

A few years ago, I would have summarized our impact by describing the career trajectories of participants as I just described, but we have increasingly realized that we've had an impact that we didn't originally intend. It's not just our students who learn, our speakers learn as well. Nearly everyone who has ever met with the group has expressed interest in speaking with the program again. This is partially because we try to create a fun atmosphere, but sometimes it is also because the speakers learn a great deal from the process. When I ask speakers why they are willing to meet with us year after year I often get an answer that goes something like: "Your students force me to think about things I don't normally think about and ask me questions I rarely am asked to answer." In many ways I think the interactions with students give our speakers a chance to also break out of their everyday routine and think about the big picture.

After our 2024 program one of our speakers reflected:

> I was able to gain insights into the research projects on multiple transformative technologies and learn about the perspectives of very qualified students on some of the policy questions currently being asked around diplomatic and policy circles. The dialogue with the group provided opportunities to share ideas and brainstorm about approaches to improve the current policy options in areas such as Artificial Intelligence.

I have always worked to create a space where the student participants could learn what they were most interested in learning. It has only recently occurred to me how much our speakers could learn from the student questions. The impact on our speakers isn't likely to fundamentally transform their work, but it's good to know that everyone walks away enlightened when you have honest conversations.

Learning from the SOtL Approach

One of the frustrations over the years with Science Outside the Lab is that it is difficult to scale. It is incredibly labor intensive, requires a significant amount of funding to run, and consumes scarce resources like the time donated by policymakers to meet with our participants. All of this work ultimately helps only a handful of people as we have found that the program works best with cohorts of roughly 15 participants. At one point we were running as many as six programs a summer, but even then we weren't even able to give 100 students a year the experience. So, unfortunately, we can't simply argue that everyone should take this program. But we do believe there are a number of important lessons that we've learned that can influence and inform other education programs. SOtL is different, but not necessarily unique. There are a number of ways to capture some aspects of it that would benefit other programs and courses.

The first reason we think the program succeeds is the engagement with practitioners. Students don't always believe what professors tell them. But practitioners—those who are working in the real world on real topics—are eminently believable to most students. Many students dismiss professors who talk about engineering ethics, but when someone with an engineering PhD working outside of academia starts talking about the impact that their decisions have made or could make, students listen. Federal science policy professionals are not the only practitioners who work at the interface of science, technology, and society. Practicing professional engineers, local EPA officials, building code inspectors, and many, many others must think about science, technology, and values on a daily basis in their jobs. They can be enrolled to help bring ethical curriculum alive even in a classroom.

Secondly, the program is successful because it addresses a topic of interest to the students. No one is forced to participate in Science Outside the Lab. Even in the earliest days, the students who took the time to apply and participate in the program were drawn to at least some aspect of it. When students are interested they are willing to dedicate significantly more time, emotion, and brain space to it. We have found that federal science and technology policy has been of interest to a number of students. But there are a variety of topics, practices, and professionals that exist at the boundary between technology and values that could be appealing to students and give them a window into technology and values that could benefit them in many ways beyond the subject matter.

In a similar vein, allowing students to direct their own learning can be incredibly valuable. As mentioned earlier, there is a wide array of things that the students don't know. I could decide what I think is most important and try to guide them to specific learning goals. But I've found that when I give them the freedom to follow their own curiosity they not only learn far more, they uncover things that I never expected myself. My role thus becomes training them to ask good questions, giving them access to people who can answer those questions, and creating a space to create where productive conversation can thrive. It can be a bit unnerving for a teacher to give up so much control, but with a strong background in the topic and trust in the speakers, the results for the students can be so powerful that I've learned to let go and trust them.

And finally, place makes a huge impact. Place is not addressed in pedagogy often enough—usually because we start with the constraints of a classroom, schedule, etc. Earlier in this chapter I described the importance of giving students the chance to explore the geographies of power of Washington, DC. That's not something that can be replicated just anywhere. But it is possible to use place to help students both focus and learn in other contexts.

For instance, one of the reasons we think SOtL succeeds is because the program gets students out of their lab. This seems trivial since coursework also gets them out of their lab. But the SOtL program gets them far enough away from their lab that they get a mental, emotional, and intellectual break from it. There is a huge temptation for PhD students to sneak back to their labs "just to get one quick thing done." Getting them far enough away that this is difficult can be incredibly valuable because when you are trying to help someone see the world in a new way, it is useful to keep them from turning back to their comfort zone and their routine.

When students embark on the program we strongly urge them to wrap up as many of their projects as possible and put an "out of office" on their e-mail because they'll be so busy with the program that they won't have time for much else. Some students manage to get a couple hours of work done at night, but for the most part they commit to the program and in doing so they are able to truly focus without their mind wandering to whether they closed the lab freezer or worry about their next experiment.

We have also found that because all the participants are together for at least eight hours a day and then in the same dorm or hotel at night, the sense of a cohort is strengthened. Students get to know each other inside the sessions and out and they begin to both build trust in each other and learn from each other. We find that the conversations last long after our meetings end. Even the movements from one meeting to another are crucial. The Metro ride from the NSF to OSTP gives students a chance to compare notes and reflect on what they're feeling. They congratulate each other on the questions they asked, share things that others might have missed, and develop ideas that they want to explore in the future. Many of these relationships and conversations last long after we all depart DC as well. All of these factors strengthen the learning that is happening and help individuals better understand what it means to them. Building cohorts of students who learn from and trust each other can significantly enhance the learning process.

Conclusion

When Ron and I were working redeveloping engineering ethics curricula there was one textbook that we thought very highly of—Mike Martin and Roland Schinzinger's *Ethics in Engineering*.[11] In particular we were captivated by their idea of framing "engineering as "social experimentation." We found their approach to be an intriguing blend of engineering ethics and STS. Their goal was to debunk the idea that engineers could both foresee and control the ways in which technology plays out after it leaves their labs. They sought to illustrate the complexity that exists in our sociotechnical world to both humble engineers and encourage them to take steps beyond the lab to ensure that their work results in the better world they hoped for.

The Science Outside the Lab program has proven to be a master class in understanding "engineering as social experimentation." The participants in the program not only learn about complexity and the ways in which technology, science, and values are intertwined. They also get to meet dozens of people who are carrying out the ethical mission that Martin and Schinzinger were pushing for. The policymakers we meet are daily working to guide science and engineering and their outputs to maximize the positive impact they have on the world. NSF program managers try to fund the work that promises to have the most important outcomes

[11] Mike Martin and Roland Schinzinger, "Engineering as Social Experimentation," in *Ethics in Engineering*, *3rd ed.* ed. by Mike Martin and Roland Schinzinger (McGraw-Hill: 1996).

for a better society. EPA regulators seek to maximize the benefits of chemicals and technologies by limiting their most damaging impacts. And congressional staffers are committed to their Member's mission to use the knowledge produced by scientists and engineers to inform policies that benefit their constituents.

Armed with the experience of an immersive week or two in the Washington, DC policy world, SOtL alumni graduates have a wealth of insights and stories about how their work intersects with broader values. Many of them take this knowledge back to their labs and use it to better understand the ethical complexities of their work and guide them in their goal to increase the chances that their work yields the benefits they hope it will yield. Others find that the ability to work at the edge of science, technology, and values on a daily basis is incredibly compelling and they shift their career path to science and technology policy. Either way they are taking ethical issues in science and engineering incredibly seriously and, whether they know it or not, have learned a bit of science and technology studies. It's one thing to teach STS to engineers and scientists. It's another thing for them to care and see ways that they can apply it to their own situations.

Acknowledgments Thanks go to Ira Bennett who co-developed the early Science Outside the Lab program with me and ran many iterations and Toby Shulruff who has helped me to organize the last several programs. This work was supported by the Nanotechnology Collaborative Infrastructure Southwest (NSF Grant #2025490) and the National Nanotechnology Coordinated Infrastructure Coordinating Office (NSF grant #2100059).

Bibliography

Austin, James. "Paths to Science Policy." *Science.org*. February 28, 2003. doi:https://doi.org/10.1126/article.64251

Bernstein, Michael J., Kiera Reifschneider, Ira Bennett, and Jameson M. Wetmore. "Science outside the lab: helping graduate students in science and engineering understand the complexities of science policy." *Science and Engineering Ethics* 23, no. 3 (2017): 861–882.

Bijker, Wiebe E., Thomas P. Hughes, and Trevor J. Pinch, eds. *The Social Construction of Technological Systems: New Directions in the Sociology and History of Technology*. MIT Press, 1987.

Canary, Heather E., Joseph R. Herkert, Karin Ellison, and Jameson M. Wetmore. "Microethics and macroethics in graduate education for scientists and engineers:

Developing and assessing instructional models." *2012 ASEE Annual Conference & Exposition* (2012): 925–934.

Harris, Charles E., Michael S. Pritchard, and Michael Rabins. *Engineering Ethics: Concepts and Cases*, 1st edition, Wadsworth Publishing, 1994.

Herkert, Joseph R. "Ways of thinking about and teaching ethical problem solving: Microethics and macroethics in engineering." *Science and Engineering Ethics* 11, no. 3 (2005): 373–385.

Johnson, Deborah G. and Jameson M. Wetmore. "STS and Ethics: Implications for Engineering Ethics." In *New Handbook of Science and Technology Studies* edited by Ed Hackett, Olga Amsterdamska, Michael Lynch, and Judy Wajcman. MIT Press, 2007, 567–582.

Johnson, Deborah G. and Jameson M. Wetmore. *Technology & Society: Building our Sociotechnical Future, Second Edition.* MIT Press, 2021.

Kline, Ronald and Trevor Pinch. "Users as agents of technological change: The social construction of the automobile in the rural United States." *Technology and Culture* 37, no. 4 (1996): 763–795.

Kline, Ronald R. "Using history and sociology to teach engineering ethics." *IEEE Technology and Society Magazine* 20, no. 4 (2001): 13–20. https://doi.org/10.1109/44.974503

Kline, Ronald R. "Engineering Case Studies: Bridging Micro and Macro Ethics." *IEEE Technology and Society Magazine* 29, no. 4 (2010): 16–19. https://doi.org/10.1109/MTS.2010.939188

Lynch, William, Ronald Kline, and Jameson M. Wetmore. "Book reviews of 'Engineering Ethics: Balancing Cost, Schedule, and Risk-Lessons Learned from the Space Shuttle' and 'The Challenger Launch Decision.'" *ISIS* 89, no. 4 (1998): 761–763.

Lynch, William T. and Ronald Kline. "Engineering Practice and Engineering Ethics." *Science, Technology, & Human Values* 25, no. 2 (2000): 195–225.

MacKenzie, Donald. *Inventing accuracy: A historical sociology of nuclear missile guidance.* MIT Press, 1990.

Martin, Mike and Roland Schinzinger. "Engineering as Social Experimentation." In *Ethics in Engineering, 3rd ed*, edited by Mike Martin and Roland Schinzinger. McGraw-Hill, 1996, 80–124.

Pinch, Trevor J., and Wiebe E. Bijker. "The social construction of facts and artefacts: Or how the sociology of science and the sociology of technology might benefit each other." *Social Studies of Science* 14, no. 3 (1984): 399–441.

Wetmore, Jameson M. "Engineering with uncertainty: Monitoring air bag performance." *Science and Engineering Ethics* 14, no. 2 (2008): 201–218.

Wetmore, Jameson M. "Delegating to the Automobile: Experimenting with Automotive Restraints in the 1970s." *Technology and Culture* 56, no. 2 (2015): 440–463.

Wetmore, Jameson M., Nalini Chhetri, and Mark Henderson. "Engineering in unfamiliar places: Two case studies for those who want to change the world." *IEEE Technology and Society* 44, no. 1 (2025): 87–100.

Woodson, Thomas S., Matthew Harsh, Michael J. Bernstein, Susan Cozzens, Jameson Wetmore, and Rafael Castillo. "Teaching community engagement to engineers via a workshop approach." *Journal of Professional Issues in Engineering Education and Practice* 145, no. 4 (2019).

CHAPTER 10

Teaching and Learning Technology Through Stories: A Conversation with Ronald R. Kline

Anto Mohsin and Jennifer L. Lieberman

Before we officially dove into questions, Ron started us off with a caveat and a promise:

RK I could get carried away with stories, too, so I'll try not to do that … But I can't explain anything without a story, so …

JL We love your stories!

RK *Sometimes my kids don't love them at all. My grandkids do.*
And from our friendly conversation, a gem of truth for all of us. We are, all of us, inextricable from stories. And it is the great joy and work of scholars across technology humanities to tell the stories that help us understand the past, the present, and the possibilities of the future in salient and salubrious ways. Then after that insight and family catch up, we officially began.

A. Mohsin
Northwestern University in Qatar, Doha, Qatar
e-mail: anto.mohsin@northwestern.edu

J. L. Lieberman (✉)
Community College of Allegheny County, Pittsburgh, PA, USA
e-mail: jlieberman@ccac.edu

J. L. Lieberman, A. Mohsin (eds.), *Technological Humanities*,
https://doi.org/10.1007/978-3-032-24741-4_10

AM Let's start from the beginning. Could you please tell us a little bit about your journey to engineering, and then from engineering into the history of science, and then STS?

RK Sure, I'll try to keep this short. I grew up in this very small town in Southeastern Kansas. 500 people. My father ran a hardware store there. So, I got really interested in electronics. I mean, he knew all about mechanics, and I got interested in electronics. I went to this high school, not in the town, but the consolidated high school. There were about 200 students in my class. And only 500 people in this town. So, getting toward graduation, I'm taking electronics, and it was a high school that had a trade school—a big trade school, too. I was getting a trade school diploma to become a technician in electronics.

And somebody came up to me and said, "you're going to be valedictorian of your class here. You're not going to college?"

I said, "Nobody in my family ever went to college." It was not on my radar at all. So, I went. There seemed like one choice: electrical engineering. Because I had studied electronics. And there's one good school for that in the state. I could go to Kansas State University, which had a big engineering program. So, I get there, and in my second or third year, I fell in love with history.

JL What do you think sparked that love of history?

RK Because of this comment a professor made about Christ being born in 4 BC. Four years before He was supposed to be born. I actually wanted to change to a history major. But the Vietnam War was on, and I had a very low draft number: 63, meaning I'd be drafted. So I took a job as an engineer, for General Electric, and I had an occupational deferment during the Vietnam War. So, I stayed in engineering. And I got a job with General Electric, and I worked on a submarine missile system. For 8 years. I did this for 8 years. And Marge [Ron's wife] tells a story: We were in Scotland, and I was the rep for the computer-controlled system that would fire these missiles. Every day I would walk … there's a submarine tender [ship] in a loch [lake] in Scotland. And every day I would walk from the tender down a gangway to a submarine moored alongside it to see, you know, how the computer did on patrol, right? That computer was bigger than this room. And I came home, and I said, I just can't do this anymore. It took that experience. Because when I was on the sub, I could see into an open missile hatch. And there were live nuclear weapons down in there. They weren't dummies. Like, I'd worked all those years on these submarines in the shipyards. I had gone on sea trials. Right? Because they were converting them. But now there were these live nuclear weapons I would see every day. And one day, Marge says, "you came home and you said, 'I can't do this anymore.'"

AM What did you say you wanted to do?

RK Well, I said I wanted to go back into history, and she said, "No. You can go back into computer science." So I could get a PhD in computer science, but not history. She said she put one husband through medical school, she's not putting another one through a PhD in history. So, first I took a couple of summer classes in 1977 at [the University of Massachusetts at] Amherst, before starting what was an AI program, actually. But, I took those classes in history. History of Russia and history of the U.S., and I saw this brochure, it said *History of Science at the University of Wisconsin*. And I knew, "oh … I can do that." Because after, like, a summer's worth of grad work in the history of America … No, I knew I didn't have enough history yet. But I knew I could do history of science with my background, right? And when we got there, it was just a great program. The main programs then were Harvard, Johns Hopkins, Princeton, Wisconsin. It was up there with those other programs, right? It was a great program.

JL That actually leads into the next question. We're wondering if there were any particularly formative courses or experiences that lit you up. Or if you have a few things that made it feel great to you, or made you feel like you'd found your intellectual home.

RK Well, first of all, it was a hard transition, right? Because I didn't speak like an academic. At all. And I didn't take notes like an academic. I'm taking notes, and there are no verbs, there are arrows. I had written poetry, and I'd written short stories. But to me, there was a disconnect between that kind of writing and scholarship … I didn't know what the humanities were, really. I didn't know what an academic in the humanities was. So, I was taking my notes with no verbs, and I remember sitting next to Bill Aspray, who's another historian of science, and he said, "what are those arrows? What do they mean?" I thought it was obvious, but it was not obvious. So, it was a hard transition. But I wanted it. And I liked it. And it was a good class. There were three people in our class, me, a fellow from France, and Sue [Susan E] Lederer, who's now head of the history of medicine at Wisconsin. And Ron [Ronald Leslie] Numbers, one of my professors, had a very powerful influence on me. Sometimes, my grad students will tell me, "You have a lot of patience with me." Well, at Wisconsin, people had a lot of patience with me. Because I was making this transition. I was teaching technical writing to engineering students. I didn't have any support from the history of science department for teaching. But people had patience with me. They saw something. It was a very close department, very close-knit.

A lot of older students like me. I was 30 years old, God, I thought I was ancient. That group, their patience. It all influenced me.

AM Wonderful, wonderful! I mean, I could definitely relate. It was a challenging transition from working as an engineer back to doing a new discipline entirely, the humanities, when I started at Cornell, so I could definitely relate, Ron.

RK Yeah.

AM So, our next question to you has a lot to do with your early work. You decided to do a dissertation on Charles Steinmetz, and then that [work] turned into your first book. And then your second book, *Consumers in the Country*, analyzed attempts to modernize the countryside by introducing new technologies, right? But you're so attentive to marginalized voices, not the typical Thomas Edison or Nikola Tesla—voices that not a lot of scholars typically studied before you. What inspired you to listen attentively to these marginalized voices in the history of science and technology?

RK Yeah, when I studied rural folks, I remember a faculty member here said, "why are you studying this?"

AM Oh, really?

RK Oh yeah. That was an interesting question. I hadn't thought really about this question before in my life. Well, with the *Steinmetz* book,[1] well, I was also interested in socialism. He was marginalized in a different way, even though he's a prominent figure, he had a deformity. I didn't think of him as a marginalized figure before you asked, but he was. There's an FBI file on him for a while, too, because he was, uh, from Germany. During World War I, before the U.S. entered the war, he had written articles in favor of

[1] Kline, *Steinmetz*.

Germany. He immigrated when Bismarck had a crackdown on the socialists. So he had experienced that marginalization. I wasn't thinking about marginalization with the rural technology book, exactly, because … rural people. That's where I came from. I grew up in this small town; my father was a merchant. And when we go back to that part of the world, my wife Marge always says, "you think of yourself as a townie. She said, there are 500 people in this town." I tell her, "I was not a farmer's kid, though." Right? But I knew farmers … my father came from the farm, and his father. My grandfather ran steam threshing machines in Oklahoma, you know, around World War I, right?

My family were all FDR Democrats. And my dad always talked about … almost like a populist kind of talk. I learned later how to categorize that talk. He'd talk about corporations, oligopolies, but wouldn't use those words. "Instead, he'd say 'big guys and the little guy.'" He was always like that. So, I think that's where a lot of my interest in writing those books came from. And my mother came from a Mennonite family.

JL Interesting, I never knew that.

RK Yeah, that influenced me, too. Back when I was an engineer, I thought of my experience looking down at the nuclear weapons on the submarines and, thought, *No, I can't do this anymore.* My family, we were pacifists. That's why they left Russia, right? That's why they left Ukraine. A lot of folks, the Jews left the Ukraine, and also Mennonites left the Ukraine to come to the U.S. So, I think a lot of it was, with that background, I had to think of folks who were not the wealthiest people in society and so forth. The marginalized part is interesting. Some of it was this background. I also got interested in gender and technology around the 1990s, when Judy Wajcman's book[2] came out. It's a great book. And I started teaching grad classes on it. I started teaching Donna Haraway, too. I was about the only person in the department to teach that type of book at the time. Margaret Rossiter was there, but her work was a different kind of feminism. It was more of equity feminism. But I've always attributed my interest in that to Marge, who is a strong feminist. Marge helped me develop my strong interest in marginalized voices. But I never thought of myself like that until I had that

[2] Judy Wajcman, *Feminism Confronts Technology* (Pennsylvania State University Press, 1991).

comment from that one faculty member. I had tenure. I had a joint appointment because I was hired under a scheme called the Minnesota model [where historians of science were placed in the departments of the subject they studied]. I was in the history and philosophy of science and technology grad program, and with the engineers, where my line was in the department of Electrical Engineering. At that time, our faculty was all over the university. I was over in engineering. Bruce Lewenstein was—still is—over in communications. People who studied physics were in physics. The philosopher of science got to be in philosophy. We were lucky. But anyway, the engineers didn't like me writing that *Consumers* book.[3]

AM Why didn't your engineering colleagues like the book *Consumers in the Country*?

RK Because I was not studying engineers, even though I was studying professionals.

AM I see.

RK The professionals were the women home economists at the universities who went out and patronizingly told farm women how to live, basically. How to run their house. And the men [agricultural engineers] would go out and tell the farmers how to work on their farms. So that book was a bit of the inverse of the *Steinmetz* book. I remember walking up Triphammer Road, and thinking, "I should write something about where I came from." It was a clear idea to do that. But I didn't realize what that meant ... not writing about elites.

AM What about the resistance element in the story, in *Consumers in the Country*?

RK So, when Trevor Pinch and I wrote the car paper for *Technology and Culture*. I had just finished the *Steinmetz* book, and there was a car conference in Norway. He said, "Let's go." I had mentioned SCOT (social construction of technology) in the book ... but you really have to look through the *Steinmetz* book to find it. But Trev said, "You wrote about Steinmetz's electric vehicle. How about we go there?" I said, "How about I do this instead?" And, this is the old days. I sent him a photograph—that picture on the cover of the *Consumers* book, of a Mennonite family. The guy sitting there,

[3] Kline, *Consumers in the Country*.

jacked up the rear wheel of the car, and that's his wife running the washer belted to the wheel. And Trevor sees this picture and says, "I'm coming over." So, he came over and we did that. So, that was part of the inspiration. But I think I was trying to answer your question about writing about users.

Well, I knew I wanted to write about users from the very beginning with that book—about how people actually use technology. And, I think I said a little about this in my speech for the Leonardo da Vinci Prize.[4] I lived in a community where my dad sold appliances, right? So, he was part of the modernizing network. Right? So I knew from growing up there, reading about rural life, working there, I knew all about that network out there.

But I knew, as well, the farm people—my father, too—we lived on 5 acres. Even though we're in town, we're on 5 acres just outside the city limits. My father, he made do. He was a tinkerer. He made do, right? Even though he sold those appliances, right? He was part of that culture of making do with what you had.

It was part of a culture where people made do ... because they came out of the Great Depression, right? And they had an environment where they could make all these alterations to this technology, right? But if they wanted it, it wasn't like they- wanted the life depicted in Stewart Brand's *Whole Earth Catalog*, the back-to-the land movement at all.

I didn't live on one of those communes, even though I knew some people who did.

But, where I'm from, there was value in getting away from all that. Value for the people ... I should have written about this in the book. The people I studied were doing it to make do with what they had, right? To make a better life for themselves, but they wanted to ... the metaphor I used was they wove urban technology into their lives, right? They wove it into the fabric of their lives. I don't know where a lot of that came from, but I know this is a post-hoc realization.

JL Sometimes we see more clearly what we meant to say, or, further studies bring ideas into a sharper focus. Like, the *Steinmetz* book reads differently after you read your entire body of work.

RK Huh, that's interesting.

[4] Kline, "Humans and Machines."

JL And I want to pull on a thread from what you were saying before, too, because you mentioned how Marge's feminism influenced your perspective. And you talked about Trevor coming over to talk about the car paper. I wonder if you'd like to say anything about mentorship or community or partnership, and the role those connections played in your work.

RK In that period, we were hired into a graduate program. People with backgrounds in history, history of science, sociology, sociology of science, philosophy, philosophy of science. And I think being in a graduate program makes a big difference, because you work together. I mean, it's not like you have ten members on a PhD committee, but whether I'm on somebody's committee or not, I knew … we all knew their work. That was a part of the way the Science & Technology Studies Program was set up. But it was also a function of who we were. Here, let me give an example. One time, Wiebe Bijker came to visit us, and he was studying S&TS departments throughout the U.S., and Europe. And he said, what accounts for the success of S&TS progam at Cornell? He kept getting various answers, and I said, look, "Those answers are fine, but how it works in practice for me is working in a [PhD] committee meeting. And you've got three people, often, one trained in sociology, one in history, maybe one in public policy. And we all work together, right?" Now, who knows how that happened? But it became kind of a … community value. When we were just starting, we didn't know if our students were going to get jobs. Until Pablo Boczkowski got a tenure-track job we didn't know somebody with a Ph.D. in S&TS was going to get a job. Right? So we were working together as a community.

AM Interesting!

RK I mean, like, there were 10 or 12 faculty members; probably at least 6 or 7 of us felt like that. And also … thankfully, Trevor was trained in British Sociology, which has an empirical bent, which means, he recognized history's value. And he was *Trevor*, too, which made a big difference. He and I had a strong friendship. There were a lot of friendships in the department before people would collaborate, too. So that was one aspect of our community. And we did not have a big department. My undergraduate teaching was in engineering; Bruce's was in communications. But our focus was mostly on grad teaching in S&TS. We were fortunate, as well, that the community lasted for a long time. Because both Trevor and I had other opportunities. I had one to replace Tom Hughes [at Penn], he had one to replace Robert Merton [at Columbia]. So, we both had a chance to replace major people in our fields, and we both said, "No, we're staying." And we didn't talk to each other about that, it just … it just happened. We talked about it later.

JL Well, it's interesting that you mention the role of graduate committees. As we worked on this collection, we noticed that *festschrifts* as a genre generally focus on the intellectual inheritance of a scholar's publications. But that's just a fractional part of the impact a scholar has on the field. We weren't surprised that mentorship emerged as a theme in the essays we've collected here, because you are such a warm and caring mentor. So, it was important to us and our contributors to push back against the tendency of *feschrifts* and other kinds of academic work to erase teaching. So, we wanted to ask if you had some thoughts about your time in the classroom, or working as a mentor. You already talked about being in the room with other committee members. Would you like to expand on that at all?

RK Sure, sure. I'm happiest teaching when somebody comes in the room for a reading class and it's just us talking, right? That's my happiest time. Period. And that reminds me of a story. When I was 5 or 6 years old, there was no kindergarten in my town. So I went to first grade. I was so excited, I came back home, and I told my younger sister everything that happened each day. I didn't have that feeling again until I got to Cornell. Did not have it in – gee– never really had when I was teaching at Wisconsin, but there was something about [the feeling of teaching and learning in community]… because to me, it always seemed like it was a communal thing. I wasn't teaching my sister, I was just excited about it, right? And at Cornell my classes felt like that. You know … I got that method from Ron Numbers. That was the way he would teach. Because the PhD students go in and talk with the professor, and you're not quite colleagues, but you're getting to be, right? I can still hear some of his questions.

JL Can you give an example?

RK There was this reading class. The very first book was a thick one on science in the British colonies of the Americas. And I was going on and on about the details, and Ron just stopped me, and he says, "Okay, next time you come in, I want to hear just one sentence about what this book is about." To me, that was a revelation. And then we could have a better discussion, because that would open up a conversation. The other way of talking about readings ... there's no conversation involved. It's like me taking notes with the arrows instead of verbs. I just now realized ... my chair at Wisconsin, Terry Reynolds, he would also do what Ron Numbers did, but Ron did it a bit more. They created a welcoming kind of feeling, but I could tell they appreciated who you were and your intelligence.

AM Well, I mean, you know, I wasn't the only one of your former advisees who felt that you really gave a lot of attention to us in mentoring, guiding us—in our research, in our dissertation writing, and also in teaching, I remember you served as a mentor for my first year writing seminar. And, in fact, later on, when you wanted to teach your first writing seminar, you also asked for *my* help.

RK Yes!

AM That was great. It meant a lot to me. Thank you.

RK Thank *you*! You sent me a book recommendation, about this thick [gestures to indicate a slim book]. I didn't want to assign a book for that class, but that was a good one. There was one chapter in there about removing clutter that I still remember. I would give an exercise with that reading. I would have the students go through something published and remove the clutter. They loved it.

But anyway, Wisconsin had very few grad seminars. It was mostly independent reading. But, you know, all of us, if you would interview somebody else from the Cornell department, Mike Lynch for example, who maybe teaches the same way, you'd get probably a different experience from them. So who knows why it works? Each scholar—each individual, in fact—brings her and his own style of teaching and mentoring.

And it changes over time.

AM Absolutely. But the care and attention that you gave to mentoring and teaching us ... I really appreciate it a lot.

RK Well, when students ask about why I care so much, one answer I give sometimes is that people took a lot of care with me when I was in grad school. They might not have known it at the time, but, day by day by day, they created a good experience for me.

JL We also felt that your style of mentorship really resonated with that theme in your work of the big and the little guy, both having something to contribute. I remember, during my time as your postdoctoral student, you would often say how we "stretch each other."

RK Yes, that's right.

JL It wasn't an empty vessel style of understanding the developing expert, but a more community-oriented perspective on knowledge-making that was really in keeping with your feminist values and your interest in socialism that you mentioned. And your research: it really permeates through to the classroom, and then what all of us have gone on to do with our whole lives since then.

RK You're right. Absolutely.

And that was the one thing about being in an S&TS department. I guess you could have stayed in your own area. But we really didn't. I always felt like I was being stretched, which was good.

AM Thank you so much, Ron! That's really wonderful, what you've shared with us. What are you working on and thinking about these days?

RK I'm trying to write a book on the history of AI. I've given a talk at the SSRG [the research symposium in Cornell's S&TS department] about AI as infrastructure. Not the infrastructure of AI, but AI itself as infrastructure, comparing the expert systems in the boom and bust of AI in the 1980s with the boom in AI today. This idea came out of a book I had planned to write based on a class I used to teach on the history of digitalization. AI was going to be one chapter in that project, but I just got obsessed with it. So, this is what I've been working on. I work a couple of hours a day. Maybe one hour a day. Two hours sometimes. It didn't really come together until I took that paper, and I kind of stretched it out into a book. I didn't want to write a history of AI, because there are a lot of people writing on the history of AI. There's Margaret Boden, I don't know if you've seen her book,[5] 1600 pages. She's just amazing. I wondered how anyone else could say anything new about the history of AI. It was published in 2006, but it's hard to imagine what she doesn't somehow cover. But, of course, there's all kinds of good answers to that. Because there are different ways to look at everything. So there was a history of AI summer school at [the University of] Cambridge—it was remote, because of Covid. And

[5] Margaret Boden, *Mind as Machine: A History of Cognitive Science* (Oxford University Press, 2006).

those folks were looking at colonialism and AI, and so forth. I'm still sorting out their arguments. But I can see my new book, on the bookshelf next to *The Cybernetics Moment* book[6]; I can see it now. I've been fighting the tendency to go back and do a deep dive into AI's technical details. Any time I have studied an engineering subject, I have to fight my tendency to do that. And I can give you one more example. I taught a grad class in science and technology in the Cold War. I have this chapter called "AI is a Cold War science." It's about 10,000 words, exploring what it means to be a Cold War science. In the last week or two, I've been re-reading those notebooks—my notes for those books I assigned for that class. All this reading we did on science during the Cold War. And the urge to dive deeply into that history is what I have to fight.

But it all goes back to the first question: the Cornell engineering professors urged me to write what other engineers want to hear about—but I want to open up my historical thinking somehow. So that's where I am.

JL This is also a little bit of a deviation from the question as we posed it originally, but you described yourself in the beginning of your time at Wisconsin as not really understanding what the humanities were yet. So I'm wondering two things: what do you think the humanities do now, and how do you think of your contribution to that work?

RK When I was in grad school, there was a debate between internalists and externalist historians of science. Margaret Rossiter was definitely on the externalist side, right? And she was happy to be there and proud of it. What she was working on, she was very proud of. And I just wrote a piece for her, a tribute to her, which is on the Cornell S&TS website, which I enjoyed writing. And there was a debate about the internalist approach, at Wisconsin and when I interviewed at Cornell. Well, fortunately, in an STS department, all that kind of stuff just kind of breaks down. It just does not make any sense. Because we're more than a history of science department. So, speaking to what the humanities do, I think about the different perspectives we can take on technical issues. There are all

[6] Kline, *The Cybernetics Moment.*

kinds of ways to do this work. Like, Ruha Benjamin has a book[7] about the way race is written into the code of AI algorithms, and there's a good book on the infrastructure of AI doing that too. But there could also eventually be user studies of AI from an S&TS or a history perspective. But I haven't come across. For me, I'm going to do what I did in *The Cybernetics Moment*. I have to think in images. There was a tendency then to just focus on information theory, and what stopped me from focusing on information theory was the Norbert Wiener Archives. He has letters with science fiction writers, philosophers, people in economics, mathematicians, high school students. I mean, the guy was just … he's just amazing … he's a polymath, right? His father was a linguist, and he himself had such a tortured childhood. And that's why it opened up my work. And then, all of a sudden, the book is opening up, and there's not just information theory: there's cybernetics, and newspapers, and even James Baldwin saying that cybernetics helped mark the 1950s for him. *That's* opening up. I hadn't thought about it until we started talking. But maybe this approach reflects my own experience of entering the humanities. There was an opening up. And when it came to that unproductive internalist/externalist divide, I wanted to be open *while* staying close to what my actors thought about their subject too.

AM Right, absolutely.

RK Here's the way I started doing it. I started just copying everything published in the media, about AI. That helps me. Some people start with theory, but I start with images.

I had fun re-reading a lot of that stuff from the 1990s about infrastructure studies. Interesting work by Bowker and Star.[8] In its early years, AI doesn't really exist outside the laboratory–not even the word *AI*. But all of a sudden, there are a lot of newspaper articles and businesses using the phrase *AI* in the 1980s. There are all these old debates about whether machines think, going back to the Golem. That old. And studying these media images over time is

[7] Ruha Benjamin, *Race After Technology: Abolitionist Tools for the New Jim Code* (Polity, 2019).

[8] Geoffrey C. Bowker and Susan Leigh Star, *Sorting Things Out: Classification and Its Consequences* (MIT Press, 2000).

how I relate AI to the broader history of technology, to politics and economics.

JL Since we got to the AI piece, and we've been thinking about community, I wonder are you thinking about AI in spaces like classrooms as well?

RK I don't know how to handle the user's part of this, I really don't. In the past, I'd find some older sources that actually interviewed users. That's how I did it with the *Consumer's* book, with those home economists who surveyed farm people. Or Sherry Turkle's work. Look at her three books.[9] They're great ethnographies, and great books too—all three of them. On the computer itself. In fact, there's a chapter on AI in *The Second Self.* I use that work, too, but … in order to understand how people engage with technology in the past, to me, those ethnographies are rich. Those are rich sources. When Trevor and I were writing the car paper, he was so frustrated that he could not take his video camera out to a farmer's field outside of Ithaca and film farmers jacking up their car to do the wash, because they don't do it anymore, not since the 1930s! It's an ethnographic site that no longer exists. And the students who have done ethnography with me, like Christina Dunbar-Hester, her sites are in the past, too, but only a few years ago. My historical sites don't exist anymore.

JL I know what you mean. It's ephemeral. Part of why I asked that question, I came to this interview after a meeting with students and some of my new colleagues where they were thinking through AI kind of ethically and personally. And one of the joys about bringing up S&TS to people who aren't familiar with the field has been talking about how SCOT (the social construction of technology) in particular—like your work, and Trevor and Bijker's work—invites us to look at how actors had these different choices. So, even when things feel black box to us—you know, the way people talk about AI in my communities now, they always imagine that it's universal adoption is a foregone conclusion. And I feel like there's so much power in recognizing throughout the history of technology that

[9] Sherry Turkle, *Life on the Screen: Identity in the Age of the Internet* (Simon & Schuster, 1995); Sherry Turkle, *The Second Self: Computers and the Human Spirit,* Twentieth Anniversary Edition (MIT Press, 2005); Sherry Turkle, *Alone Together: Why We Expect More from Technology and Less from Each Other* (Basic Books, 2011).

these were choices about policy and use, and when you know that's the case, then you know you can make different choices, too.

RK That's good. Well, that's part of the reason, too, right? Because anything with the S&TS approach is going to talk about the fact that there were choices, in the past and in the present. But I don't know how I'm going to talk about users in the past unless I do it through online archives. But users of AI embody such a wide range … Folks like the users of the internet, right? Some of that work is ephemeral and overwhelming. But what I can do in the infrastructure paper is to show this: look, we've been through AI booms before. What's different now? Well, there's a lot of things that are different, but there are a lot of things that are similar.

And also, I think, on the technical side, you can combat the utopian—or, really, dystopian technological narratives by understanding that history. My favorite book on that is by Meredith Broussard.[10] If I was still teaching, there's an example in there I would love to use about driverless cars. It's five or six years old, so maybe a little out of date, but at the time of her writing the book, you programmed driverless cars to avoid hitting, for example, orange traffic cones. There's a line in the program, *avoid traffic cones.* If the traffic cone's tipped over on its side? Then that program doesn't work, and you have to tell it to avoid traffic cones that are tipped over. You have to even tell the program what "tipped over" means too, because it does not understand common sense. Like Broussard, or Ruha Benjamin, I think that kind of humanities work can help to debunk … well, not *debunk*. It can help … I'll use the language of David Nye … provide a counter-narrative.[11] And those counternarratives can be powerful. … And I think some counter-narratives to AI will probably develop that are stronger than they are now. But who knows.

JL That's what we had. Do you feel that you have anything else that you wanted to say?

AM Yeah, is there anything else that you wanted to share that we haven't asked, Ron?

[10] Meredith Broussard, *Artificial Unintelligence: How Computers Misunderstand the World* (MIT Press, 2019).

[11] See David E. Nye, *America as Second Creation: Technology and Narratives of New Beginnings* (MIT Press, 2003).

RK Well, I think I said this when y'all did my retirement [gathering online]. I found it a privilege to teach grad students at Cornell. Period. ... So that last comment about using methods that work in studying older technologies when studying these new technologies seems to be coming from our field. We shouldn't have to be told that. [laughter].

JL Yeah.

RK But, I've seen it happen. Same with the internet too. Computer, internet, AI.

JL The people who have a lot of money and power behind the hype are invested in certain narratives, and that's what makes the counter-narratives so threatening to them, but so important.

RK Yeah, exactly. And that's what I want my work to do.

Bibliography

Benjamin, Ruha. *Race After Technology: Abolitionist Tools for the New Jim Code.* Polity, 2019.

Boden, Margaret. *Mind as Machine: A History of Cognitive Science.* Oxford University Press, 2006.

Bowker, Geoffrey C., and Susan Leigh Star. *Sorting Things Out: Classification and Its Consequences.* MIT Press, 2000.

Broussard, Meredith. *Artificial Unintelligence: How Computers Misunderstand the World.* MIT Press, 2019.

Kline, Ronald R. *Consumers in the Country: Technology and Social Change in Rural America.* Johns Hopkins University Press, 2000.

Kline, Ronald R. "Humans and Machines." *Technology and Culture* 58, no. 3 (2017): 835–45.

Kline, Ronald R. *Steinmetz: Engineer and Socialist.* Johns Hopkins University Press, 1992.

Kline, Ronald R. *The Cybernetics Moment.* Johns Hopkins University Press, 2015.

Nye, David E. *America as Second Creation: Technology and Narratives of New Beginnings.* MIT Press, 2003.

Turkle, Sherry. *Alone Together: Why We Expect More from Technology and Less from Each Other.* Basic Books, 2011.

Turkle, Sherry. *Life on the Screen: Identity in the Age of the Internet.* Simon & Schuster, 1995.

Turkle, Sherry. *The Second Self: Computers and the Human Spirit.* Twentieth Anniversary Edition. MIT Press, 2005.

Wajcman, Judy. *Feminism Confronts Technology.* Pennsylvania State University Press, 1991.

www.ingramcontent.com/pod-product-compliance
Ingram Content Group UK Ltd.
Pitfield, Milton Keynes, MK11 3LW, UK
UKHW020937180926
475692UK00004B/78
9783032247407